Palgrave Texts in Econometrics

General Editors: **Kerry Patterson** and **Terence Mills**

Regional Editors: **Northern Europe: Niels Haldrup; Far East: In Choi; USA: William Greene; Southern Europe and Oceania: Tommaso Proietti**

The pace of developments in econometrics has made it increasingly difficult for students and professionals alike to be aware of what is important and likely to be lasting in theory and practical applications. This series addresses important key developments within a unified framework, with individual volumes organised thematically. The series is relevant for both students and professionals who need to keep up with the econometric developments, yet is written for a wide audience with a style that is designed to make econometric concepts available to economists who are not econometricians.

Titles include:

Simon P. Burke and John Hunter
MODELLING NON-STATIONARY TIME SERIES

Michael P. Clements
EVALUATING ECONOMETRIC FORECASTS OF ECONOMIC AND FINANCIAL VARIABLES

Lesley Godfrey
BOOTSTRAP TESTS FOR REGRESSION MODELS

Terence C. Mills
MODELLING TRENDS AND CYCLES IN ECONOMIC TIME SERIES

Kerry Patterson
A PRIMER FOR UNIT ROOT TESTING

Kerry Patterson
UNIT ROOTS TESTS IN TIME SERIES VOLUME 1
Key Concepts and Problems

Kerry Patterson
UNIT ROOTS TESTS IN TIME SERIES VOLUME 2
Extensions and Developments

Terence C. Mills
ANALYSING ECONOMIC DATA
A Concise Introduction

Giuseppe Arbia
A PRIMER IN SPATIAL ECONOMETRICS
With Applications in R

Terence C. Mills
TIME SERIES ECONOMETRICS
A Concise Introduction

Palgrave Texts in Econometrics
Series Standing Order ISBN 978-1-4039-0172-9 (hardback)
             978-1-4039-0173-6 (paperback)
(*outside North America only*)

You can receive future titles in this series as they are published by placing a standing order. Please contact your bookseller or, in case of difficulty, write to us at the address below with your name and address, the title of the series and one of the ISBNs quoted above.

Customer Services Department, Macmillan Distribution Ltd, Houndmills, Basingstoke, Hampshire RG21 6XS, England

---

Time Series Econometrics
A Concise Introduction

Terence C. Mills
*Professor of Applied Statistics and Econometrics,
School of Business and Economics, Loughborough University, UK*

palgrave
macmillan

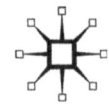

© Terence C. Mills 2015

All rights reserved. No reproduction, copy or transmission of this publication may be made without written permission.

No portion of this publication may be reproduced, copied or transmitted save with written permission or in accordance with the provisions of the Copyright, Designs and Patents Act 1988, or under the terms of any licence permitting limited copying issued by the Copyright Licensing Agency, Saffron House, 6-10 Kirby Street, London EC1N 8TS.

Any person who does any unauthorized act in relation to this publication may be liable to criminal prosecution and civil claims for damages.

The author has asserted his right to be identified as the author of this work in accordance with the Copyright, Designs and Patents Act 1988.

First published 2015 by
PALGRAVE MACMILLAN

Palgrave Macmillan in the UK is an imprint of Macmillan Publishers Limited, registered in England, company number 785998, of Houndmills, Basingstoke, Hampshire RG21 6XS.

Palgrave Macmillan in the US is a division of St Martin's Press LLC, 175 Fifth Avenue, New York, NY 10010.

Palgrave Macmillan is the global academic imprint of the above companies and has companies and representatives throughout the world.

Palgrave® and Macmillan® are registered trademarks in the United States, the United Kingdom, Europe and other countries.

ISBN: 978–1–137–52532–1

This book is printed on paper suitable for recycling and made from fully managed and sustained forest sources. Logging, pulping and manufacturing processes are expected to conform to the environmental regulations of the country of origin.

A catalogue record for this book is available from the British Library.

Library of Congress Cataloging-in-Publication Data

Mills, Terence C.
    Time series econometrics : a concise introduction / Terence C. Mills, Professor of Applied Statistics and Econometrics, Department of Economics, Loughborough University, UK.
        pages cm
    Includes index.
    ISBN 978–1–137–52532–1
    1. Econometrics. 2. Time-series analysis. I. Title.
HB139.M554 2015
330.01'51955—dc23   2015014924

Contents

List of Figures vi

List of Tables viii

1. Introduction — 1
2. Modelling Stationary Time Series: the ARMA Approach — 5
3. Non-stationary Time Series: Differencing and ARIMA Modelling — 41
4. Unit Roots and Related Topics — 58
5. Modelling Volatility using GARCH Processes — 72
6. Forecasting with Univariate Models — 85
7. Modelling Multivariate Time Series: Vector Autoregressions and Granger Causality — 98
8. Cointegration in Single Equations — 114
9. Cointegration in Systems of Equations — 134
10. Extensions and Developments — 147

Index 153

List of Figures

2.1	ACFs and simulations of AR(1) processes	11
2.2	Simulations of MA(1) processes	15
2.3	ACFs of various AR(2) processes	19
2.4	Simulations of various AR(2) processes	19
2.5	Simulations of MA(2) processes	24
2.6	Real *S&P 500* returns (annual 1872–2015)	30
2.7	UK interest rate spread (January 1952–December 2014)	31
3.1	Linear and quadratic trends	43
3.2	Explosive AR(1) model	45
3.3	Random walks	47
3.4	'Second difference' model	50
3.5	'Second difference with drift' model	51
3.6	$/£ exchange rate (January 1973–December 2014)	54
4.1	Limiting Distribution of τ	61
4.2	Limiting distribution of τ_μ	62
4.3	Limiting distribution of τ_t	66
4.4	*FTA All Share* index on a logarithmic scale (January 1952–December 2014)	67
4.5	UK interest rates (January 1952–December 2014)	69
5.1	First differences of the $/£ exchange rate (top panel). Conditional standard deviations from GARCH(1,1) model (bottom panel)	82
6.1	Logarithms of the *FTA All Share* index with linear trend superimposed	93
6.2	$/£ exchange rate from 2000 onwards with 2 conditional standard error bounds and forecasts out to December 2015	95
7.1	Orthogonalised impulse response functions: DRS and DR20 denote ΔRS and $\Delta R20$ respectively; 2 standard error bounds shown as dashed lines	109
7.2	Generalised impulse response functions	110
7.3	Variance decompositions	110
8.1	Simulated frequency distribution of $\hat{\beta}_{1000}$	117
8.2	Simulated frequency distribution of the *t*-ratio of $\hat{\beta}_{1000}$	118

8.3	Simulated frequency distribution of the spurious regression R^2	119
8.4	Simulated frequency distribution of the spurious regression dw	119
8.5	Simulated frequency distribution of $\hat{\beta}_{1000}$ from the cointegrated model with endogenous regressor	124
8.6	Simulated frequency distribution of the t-ratio on $\hat{\beta}_{1000}$ from the cointegrated model with endogenous regressor	125
8.7	Simulated frequency distribution of the slope coefficient from the stationary model with endogeneity	126
8.8	Simulated frequency distribution of the slope coefficient from the stationary model without endogeneity	126
8.9	Simulated frequency distribution of the t-ratio on $\hat{\beta}_{1000}$ from the cointegrated model with exogenous regressor	127
9.1	Error correction $e_t = R20_t - RS_t - 1.184$	144

List of Tables

2.1 SACF of real *S&P 500* returns and accompanying statistics 30
2.2 SACF and SPACF of the UK spread 32
2.3 SACF and SPACF of *FTA All Share* nominal returns 33
2.4 Model selection criteria for nominal returns 34
3.1 SACF and SPACF of the first difference of the UK spread 53
5.1 $/£ exchange rate: QML estimates 81
7.1 Order determination statistics for $\mathbf{y}_t = (\Delta RS_t, \Delta R20_t)'$ 103
9.1 Order determination statistics for $\mathbf{y}_t = (RS_t, R20_t)'$ 142

1
Introduction

About the book

1.1 The aim of this book is to provide an introductory treatment of time series econometrics that builds upon the basic statistical and regression techniques contained in my *Analysing Economic Data: A Concise Introduction*.[1] It is written from the perspective that the econometric analysis of economic and financial time series is of key importance to both students and practitioners of economics and should therefore be a core component of applied economics and of economic policy making. What I wrote in the introduction of *Analysing Economic Data* thus bears repeating in the present context: this book contains material that I think any serious student of economics and finance should be acquainted with if they are seeking to gain an understanding of a real functioning economy rather than having just a working knowledge of a set of academically constructed models of some abstract aspects of an artificial economy.

1.2 After this introductory chapter the basic concepts of stochastic processes, stationarity and autocorrelation are introduced in Chapter 2 and the class of autoregressive-moving average (ARMA) models are developed. How these models may be fitted to an observed time series is illustrated by way of a sequence of examples.

Many economic time series, however, are not stationary, but may often be transformed to stationarity by the simple operation of differencing. Chapter 3 examines some informal methods of dealing with non-stationary data and consequently introduces the key concept of an integrated process, of which the random walk is a special case, so leading to the class of autoregressive-integrated-moving average (ARIMA) models. As is demonstrated by way of examples, although the informal methods proposed in Chapter 3 often work well in practice, it is important that formal means of testing whether a series is integrated or not and, if it is, of testing what order of integration it might be, are available. Chapter 4 thus develops the theory and practice of testing for one or more unit roots, the presence of which is the manifestation of 'integratedness' of a time series. An alternative to differencing as a means of inducing stationarity is to detrend the series using a polynomial, typically a linear, function of time. How to distinguish between these two methods of inducing stationarity by way of generalised unit root tests and the differing implications of the two methods for the way the series reacts to shocks are also discussed in this chapter.

Up to this point we have assumed that the errors, or innovations, in the various models have constant variance. For many economic time series, particularly financial ones observed at relatively high frequencies, this assumption is untenable, for it is well known that financial markets go through periods of excessive turbulence followed by periods of calm, a phenomenon that goes under the general term 'volatility'. The manifestation of market volatility is that error variances change over time, being dependent upon past behaviour. Chapter 5 therefore introduces the class of autoregressive conditionally heteroskedastic (ARCH) processes. These are designed to incorporate volatility into models and, indeed, to provide estimates of such volatility.

An important aspect of time series modelling is to forecast future observations of the series being analysed. Chapter 6 develops a theory of forecasting for all the models introduced so far, emphasising how the properties of the forecasts depend in important ways on the model used to fit the data.

Only individual time series have been analysed so far, and hence the models have all been univariate in nature. Chapter 7 extends the analysis to consider a set of stationary time series, brought together as a vector and modelled as an autoregressive process, thus introducing the vector autoregression (VAR). With a vector of time series, the multivariate linkages between the individual series need to be investigated, so leading to the concept of Granger-causality, impulse response analysis and innovation accounting, all of which are discussed in this chapter.

Of course, assuming that the vector of time series is stationary is far too restrictive, but allowing the individual series to be integrated raises some interesting modelling issues for, as we demonstrate, it is possible for a linear combination of two or more integrated time series to be stationary, a concept known as cointegration. Cointegration is related to the idea of a dynamic equilibrium existing between two or more variables and, if it exists, it enables multivariate models to be expressed not only in terms of the usual differences of the series but also by the extent to which the series lie away from equilibrium: incorporating this 'equilibrium error' leads to the class of vector error correction models (VECMs).

Chapter 8 thus focuses on the consequences for conventional regression analysis when the variables in a regression are non-stationary, thus introducing the idea of spurious regression, before considering the implications of the variables in the regression being cointegrated. Tests for cointegration and estimation under cointegration are then discussed. Chapter 9 explicitly considers VECMs and how to test for and model cointegration within a VAR framework.

The final chapter, Chapter 10, explicitly recognises that this is only an introductory text on time series econometrics and so briefly discusses several extensions that more advanced researchers in the modelling of economic and financial time series would need to become familiar with. To keep within the remit of a 'concise introduction', however, no mention is made of the increasingly important subject of panel data econometrics, which combines time series with cross-sectional data, for which several textbooks are available.[2]

Mathematical level, focus and empirical exercises

1.3 As well as knowledge of basic statistics and econometrics, at the level provided by *Analysing Economic Data*, essentially all that is required to understand the material up to Chapter 7 is a good grounding in basic algebra, with some knowledge of solving equations and linear algebra plus some concepts of difference equations. Chapters 7 to 9, however, also require a basic knowledge of matrix algebra. Some technical material is placed in the notes that accompany each chapter, where key references, historical perspective and related discussion may also be found.[3]

Several examples using actual data, typically from the UK, are developed throughout the book. The content is thus suitable for final year undergraduate and postgraduate students of economics and finance wishing to undertake an initial foray into handling time series data.

1.4 Empirical exercises accompany most chapters. These are based on the software package *Econometric Views* (or *EViews*), now the industrial standard for econometric time series software, and illustrate how all the examples used in the book may be calculated and suggest how they might be extended. The data are available in an *EViews* workfile available for download.[4] It is assumed that readers already have a basic working knowledge of *EViews* or are prepared to obtain this knowledge via the extensive online help facility accompanying the package.[5]

1.5 A brief word on notation: as can be seen, chapter sections are denoted x.y, where x is the chapter and y is the section. This enables the latter to be cross-referenced as §x.y. Matrices and vectors are also written in bold font, upper case for matrices, lower case for vectors, the latter being regarded as column vectors unless otherwise stated: thus **A** is a matrix and **a** is a vector.

Notes

1. Terence C. Mills, *Analysing Economic Data: A Concise Introduction* (Palgrave Macmillan, 2014).
2. A very popular text is Badi H. Baltagi, *Econometric Analysis of Panel Data*, 5th edition (Wiley, 2013).
3. A convenient presentation of the matrix algebra required is Mills, *Matrix Representation of Regression Models: a Primer* (Lulu Press, 2013), chapter 2. Key references in time series econometrics are gathered together in Mills, *Time Series Econometrics* (Routledge, 2015).
4. At http://www.palgrave.com//resources/Product-Page-Downloads/M/Mills%20-%20Time%20Series%20Econometrics/Resources.zip
5. *EViews 8* is used throughout: see *EViews 8* (Quantitative Micro Software, LLC, Irving CA: www.eviews.com).

2
Modelling Stationary Time Series: the ARMA Approach

Stochastic processes, ergodicity and stationarity

2.1 When analysing a time series using formal statistical methods, it is often useful to regard the observations $(x_1, x_2, \ldots, x_T)$ on the series, which we shall denote generically as x_t, as a particular **realisation** of a stochastic process.[1] In general, a stochastic process can be described by a T-dimensional probability distribution, so that the relationship between a realisation and a stochastic process is analogous to that between the sample and population in classical statistics. Specifying the complete form of the probability distribution, however, will typically be too ambitious a task and we usually content ourselves with concentrating attention on the first and second moments: the T means

$$E(x_1), E(x_2), \ldots, E(x_T)$$

T variances

$$V(x_1), V(x_2), \ldots, V(x_T)$$

and $T(T-1)/2$ covariances

$$Cov(x_i, x_j), \quad i < j$$

If we could assume joint normality of the distribution, this set of expectations would then completely characterise the properties of the stochastic process. Such an assumption, however, is unlikely to be appropriate for every economic and financial series we might wish to analyse. If normality cannot be assumed, but the process is taken to be *linear*, in the sense that the current value of the process is generated by a linear combination of previous values of the process itself and current and past values of any other related processes, then again this set of expectations would capture its major properties. In either case, however, it will be impossible to infer all the values of the first and second moments from just one realisation of the process, since there are only T observations but $T+T(T+1)/2$ unknown parameters. Hence further simplifying assumptions must be made to reduce the number of unknown parameters to more manageable proportions.

2.2 We should emphasise that the procedure of using a single realisation to infer the unknown parameters of a joint probability distribution is only valid if the process is **ergodic**, which roughly means that the sample moments for finite stretches of the realisation approach their population counterparts as the length of the realisation becomes infinite. Since it is very difficult to test for ergodicity using just (part of) a single realisation, it will be assumed from now on that all time series have this property.[2]

2.3 One important simplifying assumption is that of **stationarity**, which requires the process to be in a particular state of 'statistical equilibrium'. A stochastic process is said to be **strictly stationary** if its properties are unaffected by a change of time origin. In other words, the joint probability distribution at *any* set of times $t_1, t_2, \ldots, t_m$ must be the same as the joint probability distribution at times $t_{1+k}, t_{2+k}, \ldots, t_{m+k}$, where k is an arbitrary shift in time. For $m=1$, this implies that the marginal probability distributions at $t_1, t_2, \ldots$ do not depend on time, which in turn implies that, as long as $E|x_t^2| < \infty$, both the mean and variance of x_t must be constant, so that

$$E(x_1) = E(x_2) = \cdots = E(x_T) = E(x_t) = \mu$$

and

$$V(x_1) = V(x_2) = \cdots = V(x_T) = V(x_t) = \sigma_x^2$$

If $m=2$, strict stationarity implies that all bivariate distributions do not depend on t, so that all covariances are functions only of the time-shift (or lag) k, hence implying that, for all k,

$$Cov(x_1, x_{1+k}) = Cov(x_2, x_{2+k}) = \cdots = Cov(x_{T-k}, x_T) = Cov(x_t, x_{t-k})$$

Consequently, we may define the *autocovariances* and *autocorrelations* as

$$\gamma_k = Cov(x_t, x_{t-k}) = E((x_t - \mu)(x_{t-k} - \mu))$$

and

$$\rho_k = \frac{Cov(x_t, x_{t-k})}{(V(x_t) \cdot V(x_{t-k}))^{1/2}} = \frac{\gamma_k}{\gamma_0}$$

respectively, both of which depend only on the lag k. Since these conditions apply just to the first- and second-order moments of the process, this is known as *second-order* or *weak stationarity* (and sometimes *covariance stationarity* or *stationarity in the wide sense*).

While strict stationarity (with finite second moments) thus implies weak stationarity, the converse does not hold, for it is possible for a process to be weakly stationary but *not* strictly stationary. This would be the case if higher moments, such as $E(x_t^3)$, were functions of time and an important example of this is considered in Chapter 5. If, however, joint normality could be assumed, so that the distribution was entirely characterised by the first two moments, weak stationarity would indeed imply strict stationarity.

2.4 The autocorrelations considered as a function of k are referred to as the *autocorrelation function* (ACF). Note that since

$$\gamma_k = Cov(x_t, x_{t-k}) = Cov(x_{t-k}, x_t) = Cov(x_t, x_{t+k}) = \gamma_{-k}$$

it follows that $\rho_k = \rho_{-k}$ and so only the positive half of the ACF is usually given. The ACF plays a major role in modelling dependencies among observations since it characterises, along with the process mean $\mu = E(x_t)$ and variance $\sigma_x^2 = \gamma_0 = V(x_t)$, the stationary stochastic process describing the evolution of x_t. It therefore indicates, by measuring the extent to which one value of the process is correlated with previous values, the length and strength of the 'memory' of the process.

Wold's decomposition and autocorrelation

2.5 A fundamental theorem in time series analysis, known as **Wold's decomposition**, states that every weakly stationary, purely non-deterministic, stochastic process $(x_t - \mu)$ can be written as a linear combination (or linear **filter**) of a sequence of uncorrelated random variables.[3] By purely non-deterministic we mean that any linearly deterministic components have been subtracted from $(x_t - \mu)$. Such a component is one that can be perfectly predicted from past values of itself and examples commonly found are a (constant) mean, as is implied by writing the process as $(x_t - \mu)$, periodic sequences (for example, sine and cosine functions), and polynomial or exponential sequences in t.

This linear filter representation is given by

$$x_t - \mu = a_t + \psi_1 a_{t-1} + \psi_2 a_{t-2} + \ldots = \sum_{j=0}^{\infty} \psi_j a_{t-j} \quad \psi_0 = 1 \quad (2.1)$$

The a_t, $t = 0, \pm 1, \pm 2, \cdots$ are a sequence of uncorrelated random variables, often known as **innovations**, drawn from a fixed distribution with

$$E(a_t) = 0 \quad V(a_t) = E(a_t^2) = \sigma^2 < \infty$$

and

$$Cov(a_t, a_{t-k}) = E(a_t a_{t-k}) = 0, \quad \text{for all} \quad k \neq 0$$

We will refer to such a sequence as a **white noise** process, occasionally denoting the innovations as $a_t \sim WN(0, \sigma^2)$.[4] The coefficients (possibly infinite in number) in the linear filter are known as **ψ-weights**.

2.6 We can easily show that the model (2.1) leads to autocorrelation in x_t. From this equation it follows that

$$E(x_t) = \mu$$

and

$$\begin{aligned}
\gamma_0 &= V(x_t) = E(x_t - \mu)^2 \\
&= E(a_t + \psi_1 a_{t-1} + \psi_2 a_{t-2} + \ldots)^2 \\
&= E(a_t^2) + \psi_1^2 E(a_{t-1}^2) + \psi_2^2 E(a_{t-2}^2) + \ldots \\
&= \sigma^2 + \psi_1^2 \sigma^2 + \psi_2^2 \sigma^2 + \ldots \\
&= \sigma^2 \sum_{j=0}^{\infty} \psi_j^2
\end{aligned}$$

by using the result that $E(a_{t-i} a_{t-j}) = 0$ for $i \neq j$. Now

$$\begin{aligned}
\gamma_k &= E(x_t - \mu)(x_{t-k} - \mu) \\
&= E(a_t + \psi_1 a_{t-1} + \cdots + \psi_k a_{t-k} + \cdots)(a_{t-k} + \psi_1 a_{t-k-1} + \cdots) \\
&= \sigma^2 (1 \cdot \psi_k + \psi_1 \psi_{k+1} + \psi_2 \psi_{k+2} + \cdots) \\
&= \sigma^2 \sum_{j=0}^{\infty} \psi_j \psi_{j+k}
\end{aligned}$$

and this implies

$$\rho_k = \frac{\sum_{j=0}^{\infty} \psi_j \psi_{j+k}}{\sum_{j=0}^{\infty} \psi_j^2}$$

If the number of ψ-weights in (2.1) is infinite, we have to assume that the weights are absolutely summable, so that $\sum_{j=0}^{\infty} |\psi_j| < \infty$, in which case the linear filter representation is said to *converge*. This condition can be shown to be equivalent to assuming that x_t is stationary, and guarantees that all moments exist and are independent of time, in particular that the variance of x_t, γ_0, is finite.

First-order autoregressive processes

2.7 Although equation (2.1) may appear complicated, many realistic models result from particular choices of the ψ-weights. Taking $\mu=0$ without loss of generality, choosing $\psi_j = \phi^j$ allows (2.1) to be written

$$\begin{aligned} x_t &= a_t + \phi a_{t-1} + \phi^2 a_{t-2} + \ldots \\ &= a_t + \phi(a_{t-1} + \phi a_{t-2} + \ldots) \\ &= \phi x_{t-1} + a_t \end{aligned}$$

or

$$x_t - \phi x_{t-1} = a_t \qquad (2.2)$$

This is known as a *first-order autoregressive* process, often given the acronym AR(1).[5]

2.8 The **backshift** (or **lag**) **operator** B is now introduced for notational convenience. This shifts time one step back, so that

$$B x_t \equiv x_{t-1}$$

and, in general, $B^m x_t \equiv x_{t-m}$, noting that $B^m \mu \equiv \mu$. The lag operator allows (possibly infinite) distributed lags to be written in a very concise way. For example, by using this notation the AR(1) process can be written as

$$(1 - \phi B) x_t = a_t$$

so that

$$\begin{aligned} x_t &= (1-\phi B)^{-1} a_t = (1 + \phi B + \phi^2 B^2 + \ldots) a_t \\ &= a_t + \phi a_{t-1} + \phi^2 a_{t-2} + \ldots \end{aligned} \qquad (2.3)$$

This linear filter representation will converge as long as $|\phi| < 1$, which is therefore the stationarity condition.

2.9 We can now deduce the ACF of an AR(1) process. Multiplying both sides of (2.2) by x_{t-k}, $k > 0$, and taking expectations yields

$$\gamma_k - \phi \gamma_{k-1} = E(a_t x_{t-k}). \qquad (2.4)$$

From (2.3), $a_t x_{t-k} = \sum_{i=0}^{\infty} \phi^i a_t a_{t-k-i}$. As a_t is white noise, any term in $a_t a_{t-k-i}$ has zero expectation if $k+i > 0$. Thus (2.4) simplifies to

$$\gamma_k = \phi \gamma_{k-1} \quad \text{for all} \quad k > 0$$

and, consequently, $\gamma_k = \phi^k \gamma_0$. An AR(1) process therefore has an ACF given by $\rho_k = \phi^k$. Thus if $\phi > 0$ the ACF decays exponentially to zero, while if $\phi < 0$ the ACF decays in an oscillatory pattern, both decays being slow if ϕ is close to the non-stationary boundaries of +1 and −1.

2.10 The ACFs for two AR(1) processes with (a) $\phi = 0.5$ and (b) $\phi = -0.5$ are shown in Figure 2.1, along with generated data from the two processes with a_t assumed to be normally and independently distributed with $\sigma^2 = 25$, denoted $a_t \sim NID(0,25)$, and with starting value $x_0 = 0$ (essentially a_t is normally distributed white noise, since under normality independence implies uncorrelatedness). With $\phi > 0$ adjacent values of x_t are positively correlated and the generated series has a tendency to be smooth, exhibiting runs of observations having the same sign. With $\phi < 0$, however, adjacent values have negative correlation and the generated series displays violent, rapid oscillations.

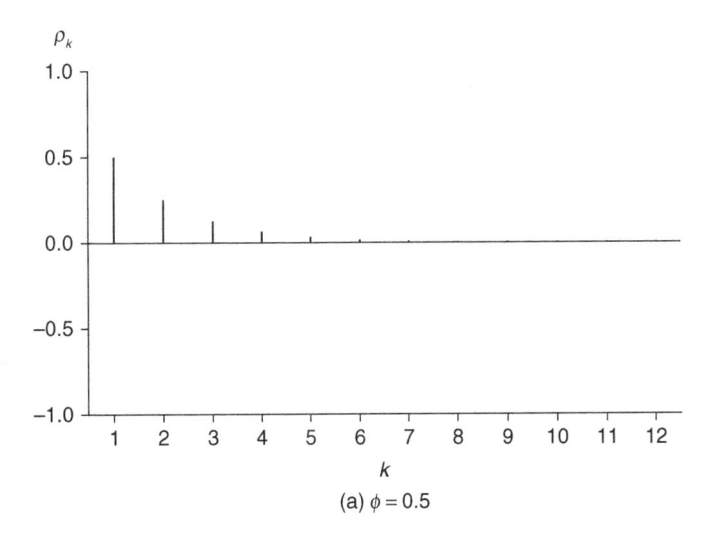

(a) $\phi = 0.5$

Figure 2.1 ACFs and simulations of AR(1) processes

12 *Time Series Econometrics*

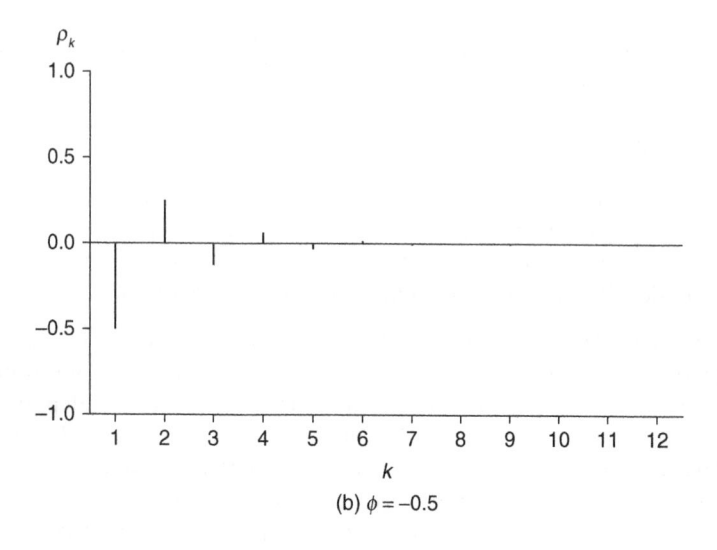

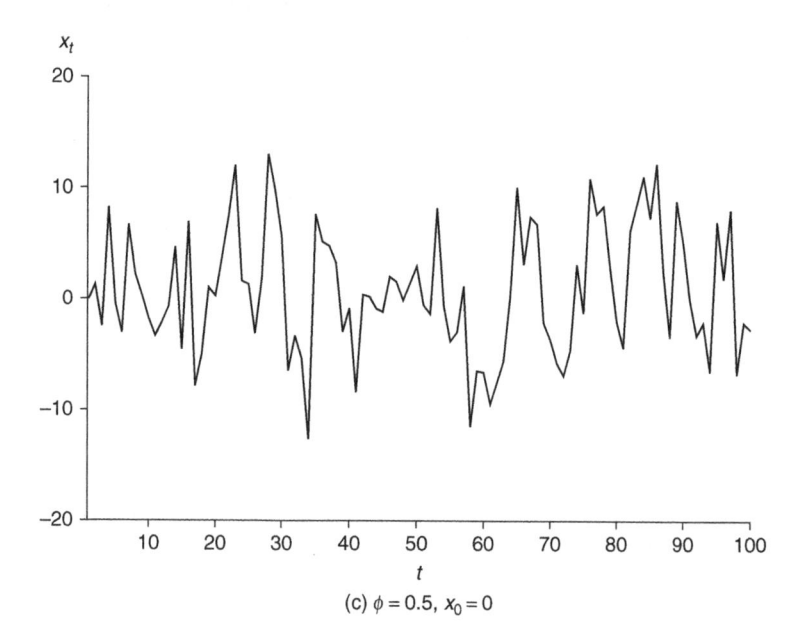

(b) $\phi = -0.5$

(c) $\phi = 0.5$, $x_0 = 0$

Figure 2.1 Continued

Modelling Stationary Time Series 13

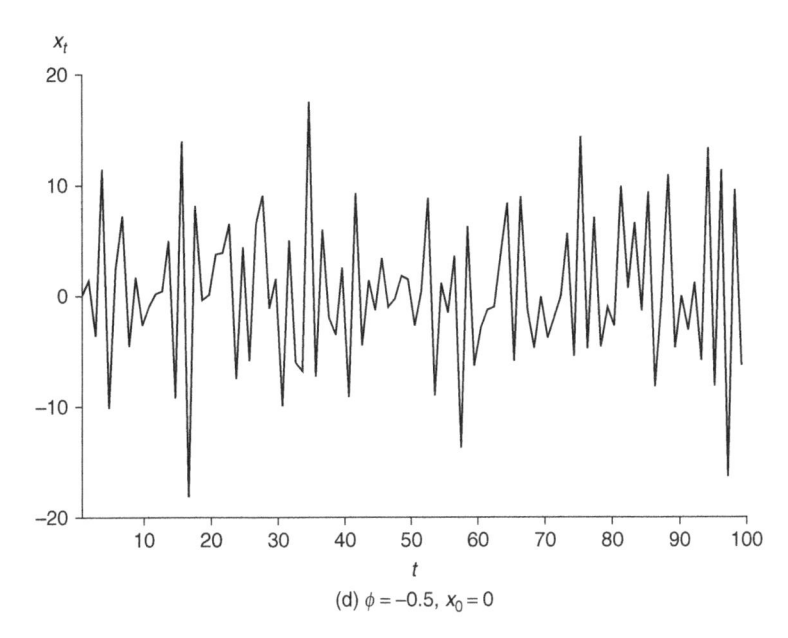

[Figure: time series plot, x_t vs t from 10 to 100]

(d) $\phi = -0.5$, $x_0 = 0$

Figure 2.1 Continued

First-order moving average processes

2.11 Now consider the model obtained by choosing $\psi_1 = -\theta$ and $\psi_j = 0$, $j \geq 2$, in (2.1):

$$x_t = a_t - \theta a_{t-1} \qquad (2.5)$$

or

$$x_t = (1 - \theta B) a_t$$

This is known as the *first-order moving average* [MA(1)] process and it follows immediately that[6]

$$\gamma_0 = \sigma^2(1 + \theta^2) \qquad \gamma_0 = -\sigma^2 \theta^2 \qquad \gamma_k = 0 \quad \text{for} \quad k > 1$$

and hence its ACF is described by

$$\rho_1 = -\frac{\theta}{1+\theta^2} \qquad \rho_k = 0 \quad \text{for} \quad k > 1$$

Thus, although observations one period apart are correlated, observations more than one period apart are not, so that the memory of the process is just one period: this 'jump' to zero autocorrelation at $k = 2$ may be contrasted with the smooth, exponential decay of the ACF of an AR(1) process.

2.12 The expression for ρ_1 can be written as the quadratic equation $\rho_1\theta^2 + \theta + \rho_1 = 0$. Since θ must be real, it follows that $|\rho_1| < 0.5$.[7] However, both θ and $1/\theta$ will satisfy this equation and thus two MA(1) processes can always be found that correspond to the same ACF.

Since any moving average model consists of a finite number of ψ-weights, all moving average models are stationary. In order to obtain a converging autoregressive representation, however, the restriction $|\theta| < 1$ must be imposed. This restriction is known as the *invertibility* condition and implies that the process can be written in terms of an infinite autoregressive representation

$$x_t = \pi_1 x_{t-1} + \pi_2 x_{t-2} + \ldots + a_t$$

where the π-*weights* converge: $\sum_{j=1}^{\infty} |\pi_j| < \infty$. In fact, the MA(1) model can be written as

$$(1 - \theta B)^{-1} x_t = a_t$$

and expanding $(1 - \theta B)^{-1}$ yields

$$(1 + \theta B + \theta^2 B^2 + \ldots) x_t = a_t.$$

The weights $\pi_j = \theta^j$ will converge if $|\theta| < 1$: in other words, if the model is invertible. This implies the reasonable assumption that the effect of past observations decreases with age.

2.13 Figure 2.2 presents plots of generated data from two MA(1) processes with (a) $\theta = 0.8$ and (b) $\theta = -0.8$, in each case again with $a_t \sim NID(0, 25)$. On comparison of these plots with those of the AR(1) processes in Figure 2.1, it is seen that realisations from the two types of processes are often quite similar (the ρ_1 values are 0.488 and 0.5, respectively, for example) suggesting that it may, on occasions, be difficult to distinguish between the two.

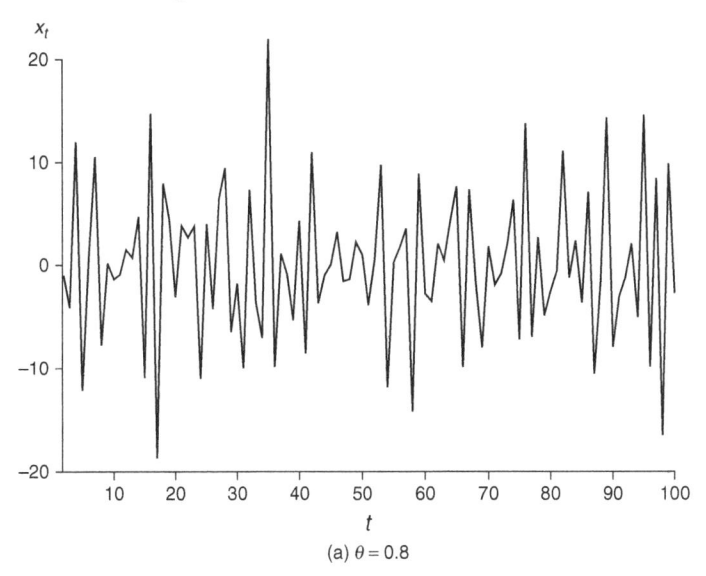

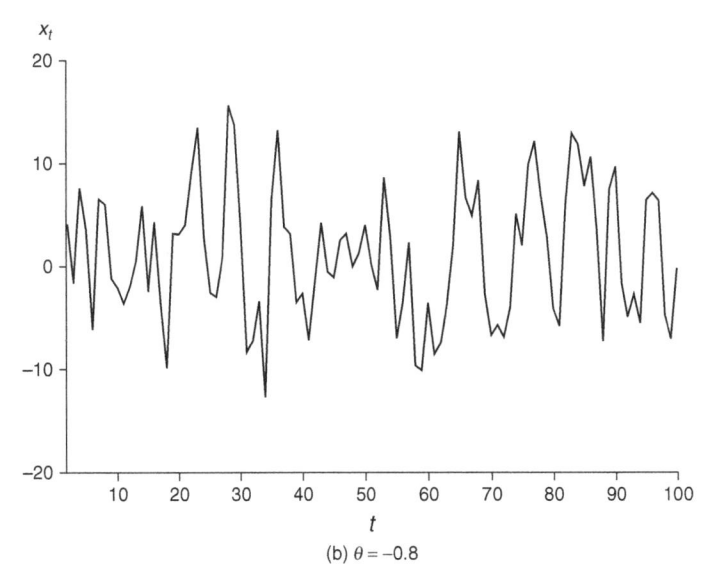

Figure 2.2 Simulations of MA(1) processes

General AR and MA processes

2.14 Extensions to the AR(1) and MA(1) models are immediate. The general autoregressive model of order p [AR(p)] can be written as

$$x_t - \phi_1 x_{t-1} - \phi_2 x_{t-2} - \ldots - \phi_p x_{t-p} = a_t$$

or

$$(1 - \phi_1 B - \phi_2 B^2 - \ldots - \phi_p B^p)x_t = \phi(B)x_t = a_t$$

The linear filter representation $x_t = \phi^{-1}(B)a_t = \psi(B)a_t$ can be obtained by equating coefficients in $\phi(B)\psi(B) = 1$.[8]

2.15 The stationarity conditions required for convergence of the ψ-weights are that the roots of the characteristic equation

$$\phi(B) = (1 - g_1 B)(1 - g_2 B)\ldots(1 - g_p B) = 0$$

are such that $|g_i| < 1$ for $i = 1, 2, \ldots, p$. The behaviour of the ACF is determined by the difference equation

$$\phi(B)\rho_k = 0 \qquad k > 0 \tag{2.6}$$

which has the solution

$$\rho_k = A_1 g_1^k + A_2 g_2^k + \ldots + A_p g_p^k$$

Since $|g_i| < 1$, the ACF is thus described by a mixture of damped exponentials (for real roots) and damped sine waves (for complex roots). As an example, consider the AR(2) process

$$(1 - \phi_1 B - \phi_2 B^2)x_t = a_t$$

with characteristic equation

$$\phi(B) = (1 - g_1 B)(1 - g_2 B) = 0$$

Modelling Stationary Time Series 17

The roots g_1 and g_2 are given by

$$g_1, g_2 = \tfrac{1}{2}\left(\phi_1 \pm \left(\phi_1^2 + 4\phi_2\right)^{1/2}\right)$$

and can both be real, or they can be a pair of complex numbers. For stationarity, it is required that the roots be such that $|g_1| < 1$ and $|g_2| < 1$ and it can be shown that these conditions imply the following set of restrictions on ϕ_1 and ϕ_2:[9]

$$\phi_1 + \phi_2 < 1 \qquad -\phi_1 + \phi_2 < 1 \qquad -1 < \phi_2 < 1$$

The roots will be complex if $\phi_1^2 + 4\phi_2 < 0$, although a necessary condition for complex roots is simply that $\phi_2 < 0$.

2.16 The behaviour of the ACF of an AR(2) process for four combinations of (ϕ_1, ϕ_2) is shown in Figure 2.3. If g_1 and g_2 are real (cases (a) and (c)), the ACF is a mixture of two damped exponentials. Depending on their sign, the autocorrelations can also damp out in an oscillatory manner. If the roots are complex (cases (b) and (d)), the ACF follows a damped sine wave. Figure 2.4 shows plots of generated time series from these four AR(2) processes, in each case with $a_t \sim NID(0,25)$. Depending on the signs of the real roots, the series may be either smooth or jagged, while complex roots tend to induce 'pseudo-periodic' behaviour.

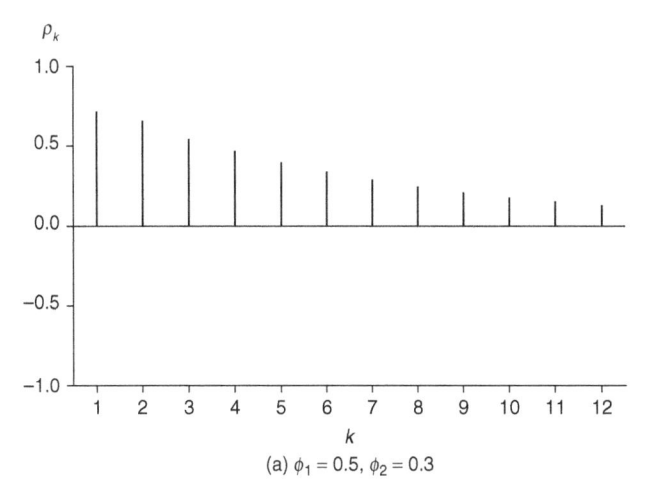

Figure 2.3 ACFs of various AR(2) processes

(b) $\phi_1 = 1$, $\phi_2 = -0.5$

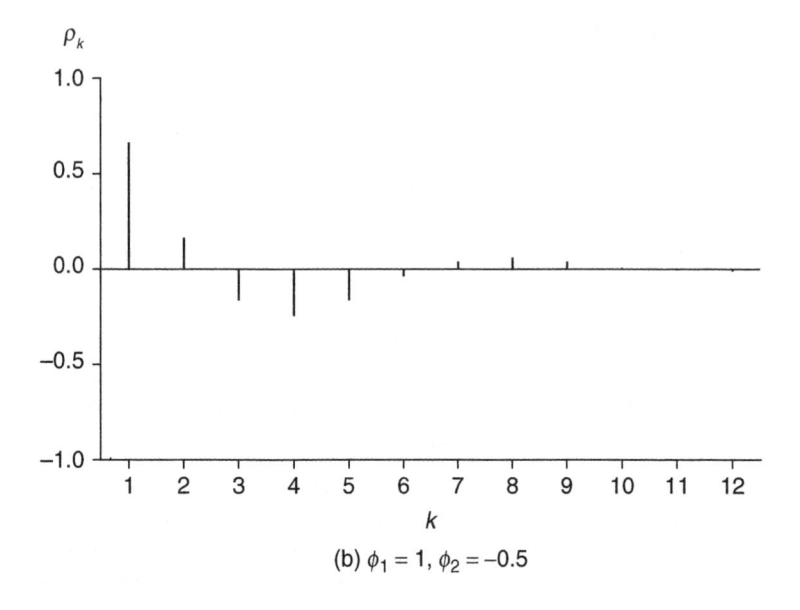

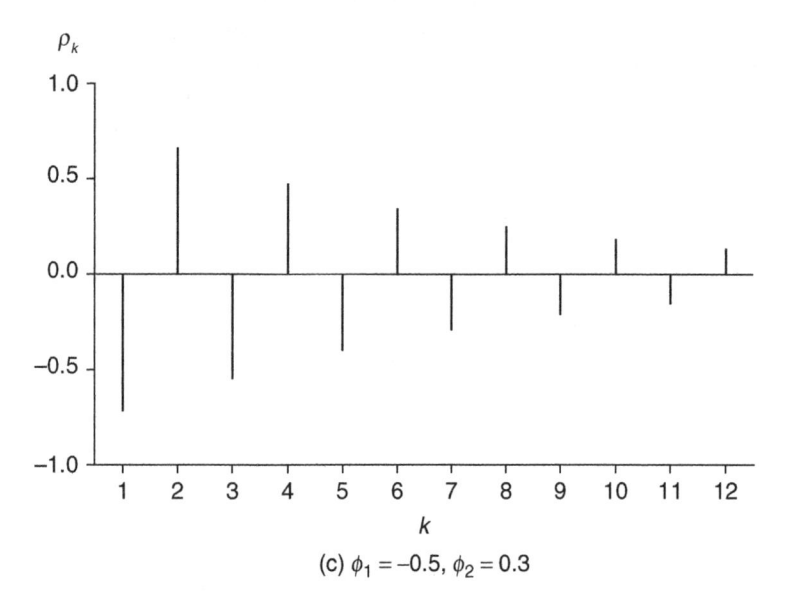

(c) $\phi_1 = -0.5$, $\phi_2 = 0.3$

Figure 2.3 Continued

(d) $\phi_1 = -0.5$, $\phi_2 = -0.3$

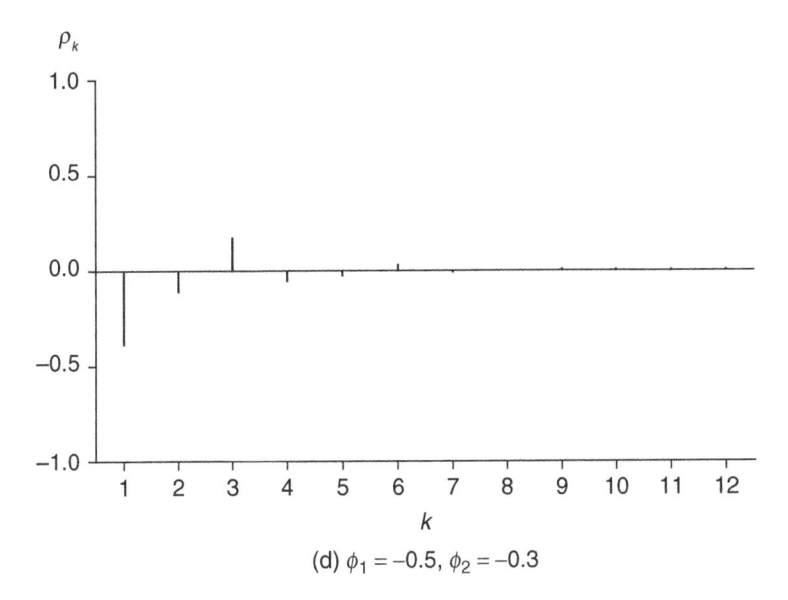

Figure 2.3 Continued

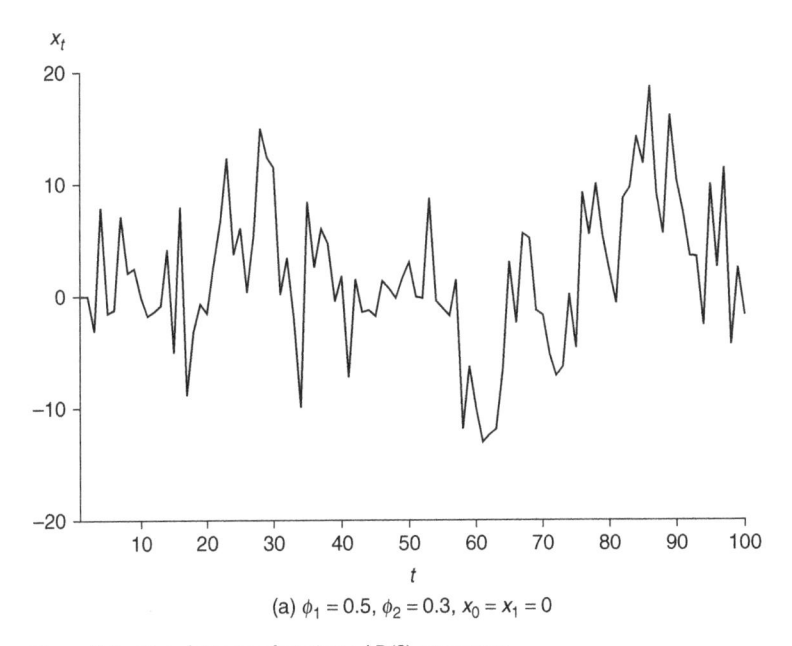

(a) $\phi_1 = 0.5$, $\phi_2 = 0.3$, $x_0 = x_1 = 0$

Figure 2.4 Simulations of various AR(2) processes

(b) $\phi_1 = 1$, $\phi_2 = -0.5$, $x_0 = x_1 = 0$

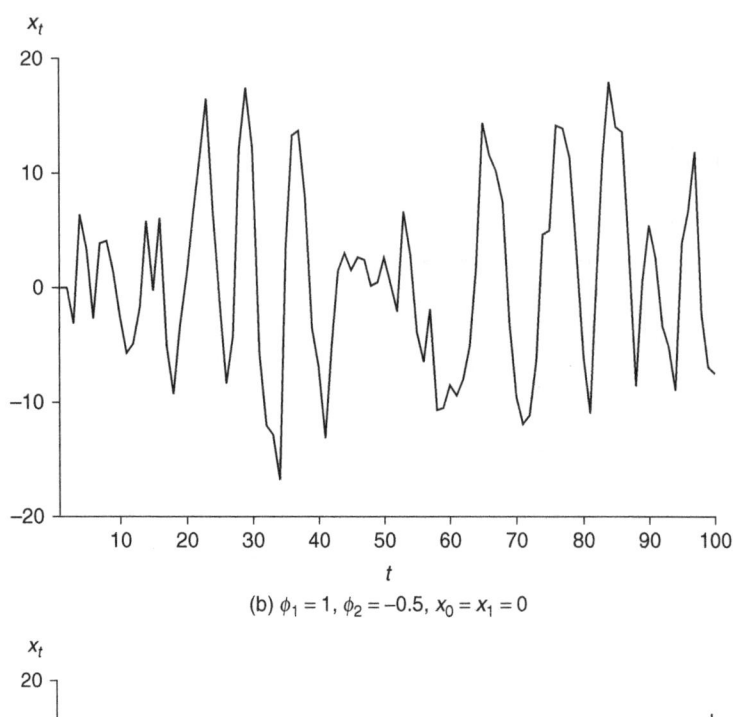

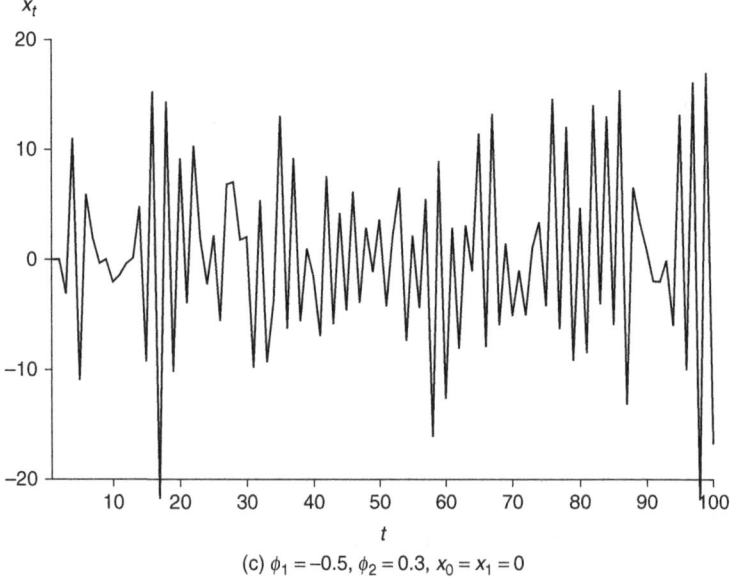

(c) $\phi_1 = -0.5$, $\phi_2 = 0.3$, $x_0 = x_1 = 0$

Figure 2.4 Continued

(d) $\phi_1 = -0.5$, $\phi_2 = -0.3$, $x_0 = x_1 = 0$

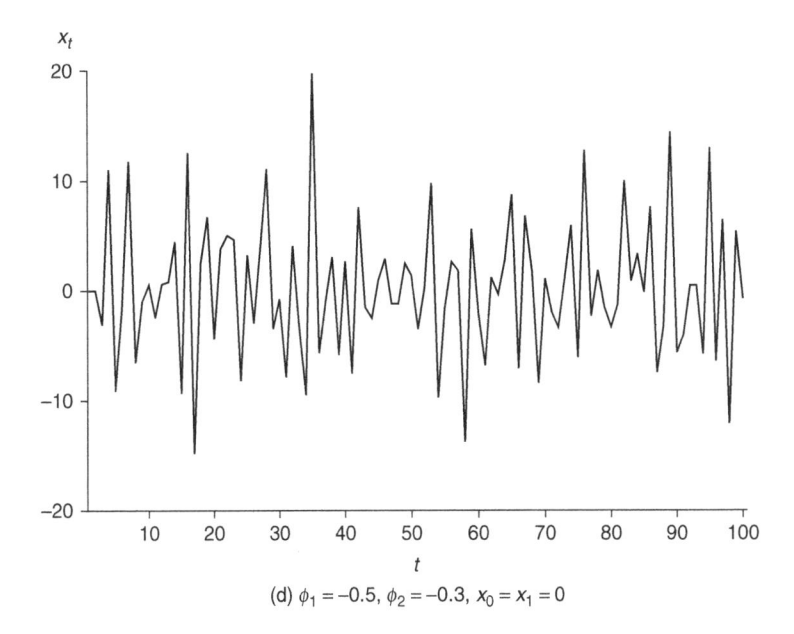

Figure 2.4 Continued

2.17 Since all AR processes have ACFs that 'damp out', it is sometimes difficult to distinguish between processes of different orders. To aid with such discrimination, we may use the ***partial autocorrelation function*** (PACF). In general, the correlation between two random variables is often due to both variables being correlated with a third. In the present context, a large portion of the correlation between x_t and x_{t-k} may be due to the correlation this pair have with the intervening lags $x_{t-1}, x_{t-2}, \ldots, x_{t-k+1}$. To adjust for this correlation, the ***partial autocorrelations*** may be calculated.

2.18 The kth partial autocorrelation is the coefficient ϕ_{kk} in the AR(k) process

$$x_t = \phi_{k1} x_{t-1} + \phi_{k2} x_{t-2} + \ldots + \phi_{kk} x_{t-k} + a_t \qquad (2.7)$$

and measures the additional correlation between x_t and x_{t-k} after adjustments have been made for the intervening lags.

In general, ϕ_{kk} can be obtained from the **Yule-Walker** equations that correspond to (2.7). These are given by the set of equations (2.6) with $p = k$ and $\phi_i = \phi_{ii}$, and solving for the last coefficient ϕ_{kk} using Cramer's Rule leads to

$$\phi_{kk} = \frac{\begin{vmatrix} 1 & \rho_1 & \cdots & \rho_{k-2} & \rho_1 \\ \rho_1 & 1 & \cdots & \rho_{k-3} & \rho_2 \\ \vdots & \vdots & \cdots & \vdots & \vdots \\ \rho_{k-1} & \rho_{k-2} & \cdots & \rho_1 & \rho_k \end{vmatrix}}{\begin{vmatrix} 1 & \rho_1 & \cdots & \rho_{k-2} & \rho_{k-1} \\ \rho_1 & 1 & \cdots & \rho_{k-3} & \rho_{k-2} \\ \vdots & \vdots & \cdots & \vdots & \vdots \\ \rho_{k-1} & \rho_{k-2} & \cdots & \rho_1 & 1 \end{vmatrix}}$$

Thus for $k = 1$, $\phi_{11} = \rho_1 = \phi$, while for $k = 2$,

$$\phi_{22} = \frac{\begin{vmatrix} 1 & \rho_1 \\ \rho_1 & \rho_2 \end{vmatrix}}{\begin{vmatrix} 1 & \rho_1 \\ \rho_1 & 1 \end{vmatrix}} = \frac{\rho_2 - \rho_1^2}{1 - \rho_1^2}$$

It then follows from the definition of ϕ_{kk} that the PACFs of AR processes follow a particular pattern:

AR(1) $\phi_{11} = \rho_1 = \phi$ $\phi_{kk} = 0$ for $k > 1$

AR(2) $\phi_{11} = \rho_1$ $\phi_{22} = \dfrac{\rho_2 - \rho_1^2}{1 - \rho_1^2}$ $\phi_{kk} = 0$ for $k > 2$

AR(p) $\phi_{11} \neq 0, \phi_{22} \neq 0, \ldots, \phi_{pp} \neq 0$ $\phi_{kk} = 0$ for $k > p$

Hence the partial autocorrelations for lags larger than the order of the process are zero. Consequently an AR(p) process is described by:

(i) an ACF that is infinite in extent and is a combination of damped exponentials and damped sine waves, and
(ii) a PACF that is zero for lags larger than p.

2.19 The general moving average of order q [MA(q)] can be written as

$$x_t = a_t - \theta_1 a_{t-1} - \ldots - \theta_q a_{t-q}$$

or

$$x_t = \left(1 - \theta_1 B - \ldots - \theta_q B^q\right) a_t = \theta(B) a_t$$

The ACF can be shown to be

$$\rho_k = \frac{-\theta_k + \theta_1 \theta_{k+1} + \ldots + \theta_{q-k} \theta_q}{1 + \theta_1^2 + \ldots + \theta_q^2} \quad k = 1, 2, \ldots, q$$

$$\rho_k = 0 \quad k > q$$

The ACF of an MA(q) process therefore cuts off after lag q: the memory of the process extends q periods, observations more than q periods apart being uncorrelated.

2.20 The weights in the AR(∞) representation $\pi(B) x_t = a_t$ are given by $\pi(B) = \theta^{-1}(B)$ and can be obtained by equating coefficients of B^j in $\pi(B)\theta(B) = 1$. For invertibility, the roots of

$$(1 - \theta_1 B - \ldots - \theta_q B^q) = (1 - h_1 B) \cdots (1 - h_q B) = 0$$

must satisfy $|h_i| < 1$ for $i = 1, 2, \ldots, q$.

2.21 Figure 2.5 presents generated series from two MA(2) processes, again using $a_t \sim NID(0,25)$. The series tend to be fairly jagged, similar to AR(2) processes with real roots of opposite signs, and, of course, such MA processes are unable to capture periodic-type behaviour.

2.22 The PACF of an MA(q) process can be shown to be infinite in extent, so that it tails off. Explicit expressions for the PACFs of MA processes are complicated but, in general, are dominated by combinations of exponential decays (for the real roots in $\theta(B)$) and/or damped sine waves (for the complex roots). Their patterns are thus very similar to the ACFs of AR processes.

(a) $\theta_1 = -0.5$, $\theta_2 = 0.3$

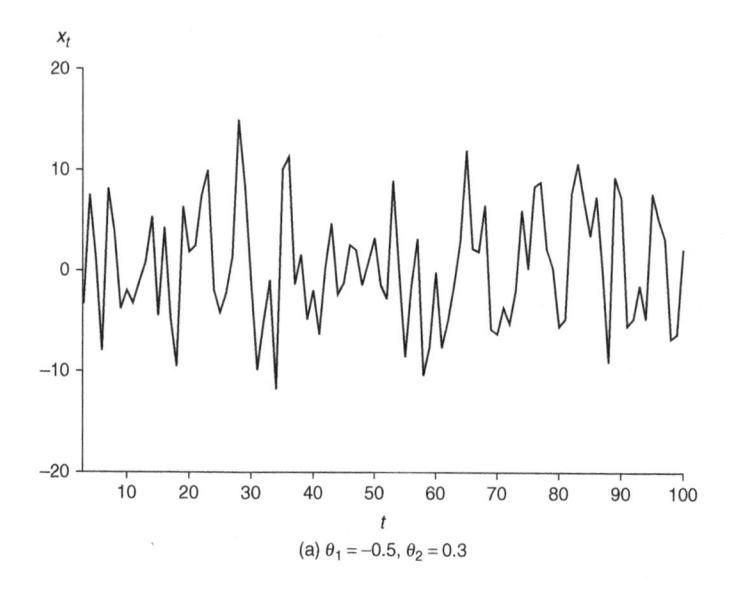

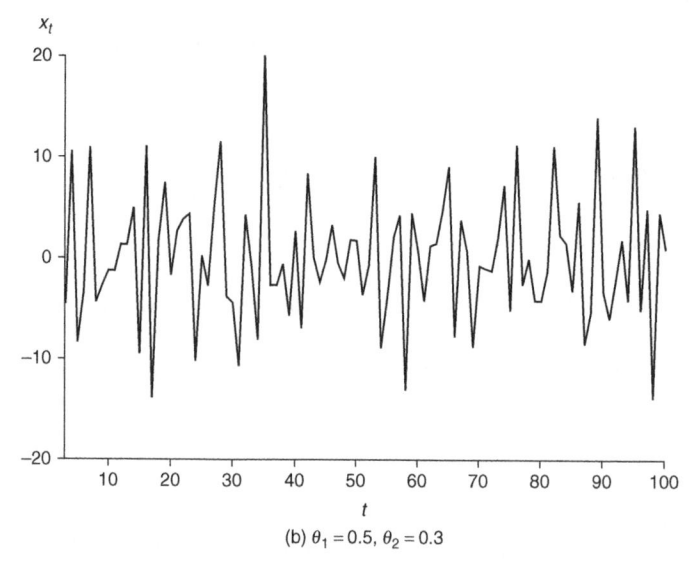

(b) $\theta_1 = 0.5$, $\theta_2 = 0.3$

Figure 2.5 Simulations of MA(2) processes

Modelling Stationary Time Series 25

Indeed, an important duality between AR and MA processes exists: while the ACF of an AR(p) process is infinite in extent, the PACF cuts off after lag p. The ACF of an MA(q) process, on the other hand, cuts off after lag q, while the PACF is infinite in extent.

Autoregressive-moving average models

2.23 We may also entertain combinations of autoregressive and moving average models. For example, consider the natural combination of the AR(1) and MA(1) models, known as the *first-order autoregressive-moving average*, or ARMA(1,1), process

$$x_t - \phi x_{t-1} = a_t - \theta a_{t-1} \tag{2.8}$$

or

$$(1 - \phi B)x_t = (1 - \theta B)a_t$$

The ψ-weights in the MA(∞) representation are given by

$$\psi(B) = \frac{(1 - \theta B)}{(1 - \phi B)}$$

so that

$$x_t = \psi(B)a_t = \left(\sum_{i=0}^{\infty} \phi^i B^i\right)(1 - \theta B)a_t = a_t + (\phi - \theta)\sum_{i=1}^{\infty} \phi^{i-1} a_{t-i} \tag{2.9}$$

Likewise, the π-weights in the AR(∞) representation are given by

$$\pi(B) = \frac{(1 - \phi B)}{(1 - \theta B)}$$

so that

$$\pi(B)x_t = \left(\sum_{i=0}^{\infty} \theta^i B^i\right)(1 - \phi B)x_t = a_t$$

or

$$x_t = (\phi - \theta)\sum_{i=1}^{\infty} \theta^{i-1} x_{t-i} + a_t$$

The ARMA(1,1) process thus leads to both moving average and autoregressive representations having an infinite number of weights. The ψ-weights converge for $|\phi|<1$ (the stationarity condition) and the π-weights converge for $|\theta|<1$ (the invertibility condition). The stationarity condition for the ARMA(1,1) model is thus the same as that for an AR(1) model.

2.24 From equation (2.9) it is clear that any product $x_{t-k}a_{t-j}$ has zero expectation if $k > j$. Thus multiplying both sides of (2.8) by x_{t-k} and taking expectations yields

$$\gamma_k = \phi \gamma_{k-1} \quad \text{for} \quad k > 1$$

whilst for $k = 0$ and $k = 1$ we obtain, respectively,

$$\gamma_0 - \phi \gamma_1 = \sigma^2 - \theta(\phi - \theta)\sigma^2$$

and

$$\gamma_1 - \phi \gamma_0 = -\theta \sigma^2$$

Eliminating σ^2 from these two equations allows the ACF of the ARMA(1,1) process to be given, after some algebraic manipulation, by

$$\rho_1 = \frac{(1-\phi\theta)(\phi-\theta)}{1+\theta^2 - 2\phi\theta}$$

and

$$\rho_k = \phi \rho_{k-1} \quad \text{for} \quad k > 1$$

The ACF of an ARMA(1,1) process is therefore similar to that of an AR(1) process, in that the autocorrelations decay exponentially at a rate ϕ. Unlike the AR(1), however, this decay starts from ρ_1 rather than from $\rho_0 = 1$. Moreover, $\rho_1 \neq 1$ and, since for typical economic and financial series both ϕ and θ will be positive with $\phi > \theta$, ρ_1 can be much less than ϕ if $\phi - \theta$ is small.

2.25 More general ARMA models are obtained by combining AR(p) and MA(q) processes:

$$x_t - \phi_1 x_{t-1} - \ldots - \phi_p x_{t-p} = a_t - \theta_1 a_{t-1} - \ldots - \theta_q a_{t-q}$$

or

$$(1 - \phi_1 B - \ldots - \phi_p B^p) x_t = (1 - \theta_1 B - \ldots - \theta_q B^q) a_t \qquad (2.10)$$

which may be written more concisely as

$$\phi(B) x_t = \theta(B) a_t$$

The resultant ARMA(p,q) process has the stationarity and invertibility conditions associated with the constituent AR(p) and MA(q) processes respectively. Its ACF will eventually follow the same pattern as that of an AR(p) process after $q - p$ initial values $\rho_1, \ldots, \rho_{q-p}$, while its PACF eventually (for $k > q - p$) behaves like that of an MA(q) process.

2.26 Throughout this development, we have assumed that the mean of the process, μ, is zero. Non-zero means are easily accommodated by replacing x_t with $x_t - \mu$ in (2.10), so that in the general case of an ARMA(p,q) process, we have

$$\phi(B)(x_t - \mu) = \theta(B) a_t$$

Noting that $\phi(B)\mu = (1 - \phi_1 - \ldots - \phi_p)\mu = \phi(1)\mu$, the model can equivalently be written as

$$\phi(B) x_t = \theta_0 + \theta(B) a_t$$

where $\theta_0 = \phi(1)\mu$ is a constant or intercept.

ARMA model building

2.27 An essential first step in fitting ARMA models to observed time series is to obtain estimates of the generally unknown parameters μ, σ_x^2 and the ρ_k. With our stationarity and (implicit) ergodicity assumptions, μ and σ_x^2 can be estimated by the sample mean and sample variance, respectively, of the realisation $(x_1, x_2, \ldots, x_T)$:

$$\bar{x} = T^{-1} \sum_{t=1}^{T} x_t$$

$$s^2 = T^{-1} \sum_{t=1}^{T} (x_t - \bar{x})^2$$

An estimate of ρ_k is then given by the lag k *sample autocorrelation*

$$r_k = \frac{\sum_{t=k+1}^{T} (x_t - \bar{x})(x_{t-k} - \bar{x})}{Ts^2} \qquad k = 1, 2, \ldots$$

the set of r_ks defining the *sample autocorrelation function* (SACF).

For independent observations drawn from a fixed distribution with finite variance ($\rho_k = 0$, for all $k \neq 0$), the variance of r_k is approximately given by T^{-1}. If, as well, T is large, $\sqrt{T} r_k$ will be approximately standard normal, so that $r_k \stackrel{a}{\sim} N(0, T^{-1})$, implying that an absolute value of r_k in excess of $2/\sqrt{T}$ may be regarded as 'significantly' different from zero. More generally, if $\rho_k = 0$ for $k > q$, the variance of r_k, for $k > q$, is

$$V(r_k) = T^{-1}\left(1 + 2\rho_1^2 + \ldots + 2\rho_q^2\right) \qquad (2.11)$$

Thus, by successively increasing the value of q and replacing the ρ_ks by their sample estimates, the variances of the sequence $r_1, r_2, \ldots, r_k$ can be estimated as T^{-1}, $T^{-1}(1 + 2r_1^2)$, $\ldots$, $T^{-1}(1 + 2r_1^2 + \ldots + 2r_{k-1}^2)$ and, of course, these will be larger, for $k > 1$, than those calculated using the simple formula T^{-1}.

2.28 The *sample partial autocorrelation function* (SPACF) is usually calculated by fitting autoregressive models of increasing order: the estimate of the last coefficient in each model is the sample partial autocorrelation, $\hat{\phi}_{kk}$.[10] If the data follow an AR(p) process then, for lags greater than p, the variance of $\hat{\phi}_{kk}$ is approximately T^{-1}, so that $\hat{\phi}_{kk} \overset{a}{\sim} N(0, T^{-1})$.

2.29 Given the r_k and $\hat{\phi}_{kk}$, along with their respective standard errors, the approach to ARMA model building proposed by George Box and Gwilym Jenkins (the Box-Jenkins approach) is essentially to match the behaviour of the SACF and SPACF of a particular time series with that of various theoretical ACFs and PACFs, picking the best match (or set of matches), estimating the unknown model parameters (the ϕ_is, θ_is and σ^2), and checking the residuals from the fitted models for any possible misspecifications.[11]

Another popular method is to select a set of models based on prior considerations of maximum settings of p and q, estimate each possible model and select that model which minimises a chosen selection criterion based on goodness of fit considerations. These model building procedures will not be discussed in detail: rather, they will be illustrated by way of a sequence of examples.

EXAMPLE 2.1 Are the returns on the *S&P 500* a 'fair game'?

An important and often analysed financial series is the real return on the annual **Standard & Poor (S&P) 500** stock index for the US. Annual observations from 1872 to 2015 ($T = 145$) are plotted in Figure 2.6 and the SACF up to $k = 12$ is given in Table 2.1. It is seen that the series appears to be stationary around a constant mean, estimated to be 3.72%. This is confirmed by the SACF and a comparison of each of the r_k with their corresponding standard errors, computed using equation (2.11), shows that none are individually significantly different from zero, thus suggesting that the series is, in fact, white noise.

We can also construct a 'portmanteau' statistic based on the complete set of r_ks. On the hypothesis that $x_t \sim WN(\mu, \sigma^2)$, then

$$Q(k) = T(T+2) \sum_{i=1}^{k} (T-i)^{-1} r_i^2 \overset{a}{\sim} \chi_k^2$$

$Q(k)$ statistics, with accompanying marginal significance levels of rejecting the null, are also reported in Table 2.1 for $k = 1, 2, \ldots, 12$ and they confirm that there is no evidence against the null hypothesis

30 Time Series Econometrics

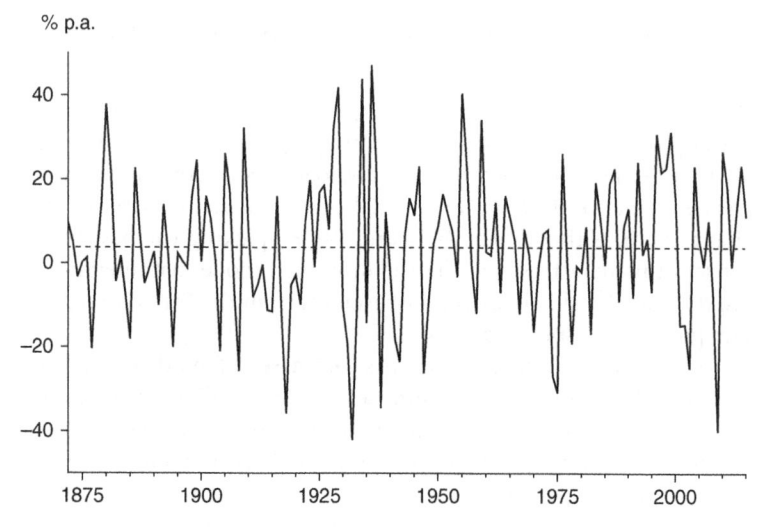

% p.a.

[Figure: line chart of real S&P 500 returns, y-axis from −40 to 40, x-axis from 1875 to 2000]

Figure 2.6 Real S&P 500 returns (annual 1872–2015)

Table 2.1 SACF of real S&P 500 returns and accompanying statistics

k	r_k	$se(r_k)$	$Q(k)$
1	0.080	0.083	0.95 [0.33]
2	−0.163	0.083	4.89 [0.09]
3	0.074	0.085	5.70 [0.13]
4	−0.087	0.086	6.83 [0.14]
5	−0.151	0.086	10.27 [0.07]
6	0.061	0.087	10.83 [0.09]
7	0.120	0.087	13.03 [0.07]
8	−0.046	0.088	13.36 [0.10]
9	−0.055	0.089	13.82 [0.13]
10	0.040	0.089	14.07 [0.17]
11	−0.034	0.090	14.25 [0.22]
12	−0.104	0.090	15.96 [0.19]

Note: Figures in [..] give $P(\chi^2_k > Q(k))$

that returns are white noise.[12] Real returns on the *S&P 500* would therefore appear to be consistent with the fair game model in which the expected return is constant, being 3.72% per annum.

EXAMPLE 2.2 Modelling the UK interest rate spread

The 'spread', the difference between long and short interest rates, is an important variable in testing the expectations hypothesis of the term structure of interest rates. Figure 2.7 shows the spread between 20 year UK gilts and 91 day Treasury bills using monthly observations for the period January 1952 to December 2014 ($T = 756$), while Table 2.2 reports the SACF and SPACF up to $k = 12$, with accompanying standard errors.

The spread is seen to be considerably smoother than one would expect if it was a realisation from a white noise process, and this is confirmed by the SACF, all of whose values are positive and significant (the accompanying portmanteau statistic is $Q(12) = 5626!$). The SPACF has both $\hat{\phi}_{11}$ and $\hat{\phi}_{22}$ significant, thus identifying an AR(2) process. Fitting such a model to the series by ordinary least squares (OLS) regression yields

$$x_t = 0.035 + 1.192\ x_{t-1} - 0.224\ x_{t-2} + \hat{a}_t \qquad \hat{\sigma} = 0.401$$
$$(0.017)\ (0.036)\phantom{\ x_{t-1}}(0.036)$$

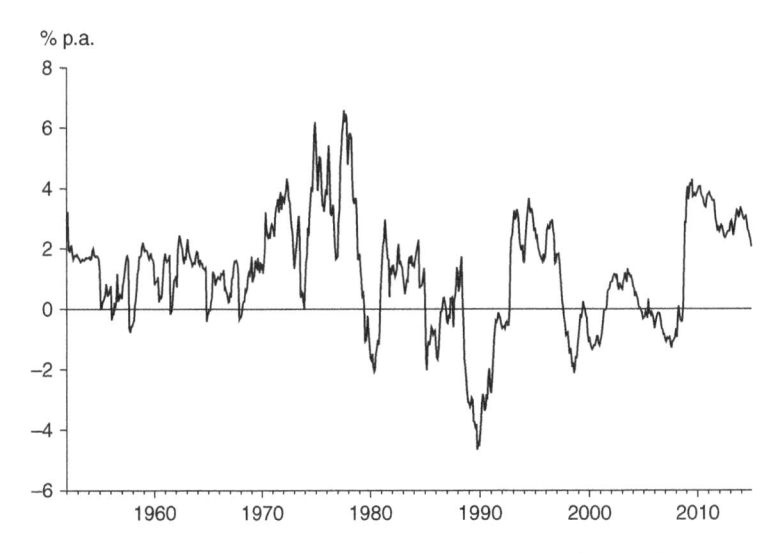

Figure 2.7 UK interest rate spread (January 1952 – December 2014)

Table 2.2 SACF and SPACF of the UK spread

k	r_k	$se(r_k)$	$\hat{\phi}_{kk}$	$se(\hat{\phi}_{kk})$
1	0.974	0.036	0.974	0.036
2	0.938	0.061	−0.220	0.036
3	0.901	0.078	0.015	0.036
4	0.864	0.090	−0.005	0.036
5	0.827	0.100	−0.047	0.036
6	0.789	0.109	−0.049	0.036
7	0.750	0.116	−0.009	0.036
8	0.712	0.122	−0.006	0.036
9	0.677	0.127	0.042	0.036
10	0.647	0.132	0.037	0.036
11	0.618	0.136	−0.007	0.036
12	0.590	0.140	−0.016	0.036

Figures in parentheses are standard errors and the intercept implies a fitted mean of $\hat{\mu} = \hat{\theta}_0/(1-\hat{\phi}_1-\hat{\phi}_2) = 1.128$ with standard error 0.467. The model can therefore be written in 'mean deviation' form as

$$x_t = 1.128 + 1.192(x_{t-1} - 1.128) - 0.224(x_{t-2} - 1.128) + \hat{a}_t$$

Since $\hat{\phi}_1 + \hat{\phi}_2 = 0.968$, $-\hat{\phi}_1 + \hat{\phi}_2 = -1.416$ and $\hat{\phi}_2 = -0.224$, the stationarity conditions associated with an AR(2) process are satisfied but, although $\hat{\phi}_2$ is negative, $\hat{\phi}_1^2 + 4\hat{\phi}_2 = 0.525$ so that the roots are real, being $g_1 = 0.96$ and $g_2 = 0.23$. The spread is thus stationary around an 'equilibrium' level of 1.128: equivalently, in equilibrium, long rates are 1.128 percentage points higher than short rates. The closeness of g_1 to unity will be discussed further in Example 4.1, but its size means that shocks that force the spread away from its equilibrium will take a long time to dissipate and hence the spread will have long departures away from this level, although as the roots are real these departures will not follow any cyclical pattern.

Having fitted an AR(2) process, it is now necessary to establish whether such a model is adequate. As a 'diagnostic check', we may examine the properties of the residuals $\hat{a}_t$. Since these are estimates of a_t, they should mimic its behaviour, so that they should behave as white noise. The portmanteau statistic Q can be used for this purpose, although the degrees of freedom must be amended: if an ARMA(p,q) process is fitted, they are reduced to $k - p - q$. With $k = 12$, our residuals yield the value $Q(12) = 7.50$, which is now asymptotically distributed as $\chi^2(10)$ and hence gives no evidence of model inadequacy. An alternative approach to assessing model adequacy is to overfit. For example, we might consider fitting an AR(3) process or, perhaps, an ARMA(2,1) to the series.[13] These yield the following pair of models

$$x_t = \underset{(0.017)}{0.036} + \underset{(0.036)}{1.194}\, x_{t-1} - \underset{(0.056)}{0.222}\, x_{t-2} - \underset{(0.036)}{0.002}\, x_{t-3} + \hat{a}_t \quad \hat{\sigma} = 0.399$$

$$x_t = \underset{(0.019)}{0.036} + \underset{(0.160)}{1.148}\, x_{t-1} - \underset{(0.156)}{0.181}\, x_{t-2} + \hat{a}_t + \underset{(0.162)}{0.046}\, \hat{a}_{t-1} \quad \hat{\sigma} = 0.401$$

The additional parameter is insignificant in both models, thus confirming the adequacy of our original choice of an AR(2) process.

EXAMPLE 2.3 Modelling returns on the FTA-All Share Index

The broadest based stock index in the UK is the **Financial Times-Actuaries (FTA) All Share**. Table 2.3 reports the SACF and SPACF (up to $k = 12$) of its nominal return calculated from monthly observations of the index from January 1952 to December 2014 ($T = 756$). The portmanteau statistic is $Q(12) = 31.3$, with a marginal significance level of 0.002, and r_1, r_9 and $\hat{\phi}_{kk}$ at lags $k = 1,2,3, 5$ and 9 are all greater than 2 standard errors. This suggests that the series is best modelled by some ARMA process of reasonably low order, although a number of models could be consistent with the behaviour shown by the SACF and SPACF.

Table 2.3 SACF and SPACF of *FTA All Share* nominal returns

k	r_k	$se(r_k)$	$\hat{\phi}_{kk}$	$se(\hat{\phi}_{kk})$
1	0.117	0.036	0.117	0.036
2	−0.062	0.036	−0.076	0.036
3	0.061	0.037	0.080	0.036
4	0.072	0.037	0.051	0.036
5	−0.067	0.037	−0.075	0.036
6	−0.046	0.037	−0.024	0.036
7	0.027	0.037	0.018	0.036
8	−0.019	0.037	−0.026	0.036
9	0.075	0.037	0.101	0.036
10	0.008	0.037	−0.021	0.036
11	−0.029	0.037	−0.022	0.036
12	0.022	0.037	0.025	0.036

In such circumstances, there are a variety of selection criteria that may be used to choose an appropriate model, of which perhaps the most popular is *Akaike's Information Criterion*, defined as

$$AIC(p,q) = \log \hat{\sigma}^2 + 2(p+q)T^{-1}$$

although a criterion that has better theoretical properties is *Schwarz's*:

$$BIC(p,q) = \log \hat{\sigma}^2 + (p+q)T^{-1} \log T.$$

The criteria are used in the following way. Upper bounds, say p_{max} and q_{max}, are set for the orders of $\phi(B)$ and $\theta(B)$ and, with $\bar{p} = \{0,1,\ldots,p_{max}\}$ and $\bar{q} = \{0,1,\ldots,q_{max}\}$, orders p_1 and q_1 are selected such that, for example,

$$AIC(p_1,q_1) = \min AIC(p,q) \qquad p \in \bar{p}, \quad q \in \bar{q}$$

with parallel strategies obviously being employed in conjunction with BIC.[14] One possible difficulty with the application of this strategy is that no specific guidelines on how to determine $\bar{p}$ and $\bar{q}$ seem to be available, although they are tacitly assumed to be sufficiently large for the range of models to contain the 'true' model, which we may denote as having orders (p_0,q_0) and which, of course, will not necessarily be the same as (p_1,q_1), the orders chosen by the criterion under consideration.

Given these alternative criteria, are there reasons for preferring one to the other? If the true orders (p_0,q_0) are contained in the set (p,q),

Table 2.4 Model selection criteria for nominal returns

	q	0	1	2	3
	p				
AIC	0	−3.107	−3.121	−3.123	−3.122
	1	−3.117	−3.122	−3.119	−3.119
	2	−3.120	−3.120	−*3.141*	−3.125
	3	−3.123	−3.121	−3.125	−3.136
BIC	0	−3.101	−*3.108*	−3.104	−3.098
	1	−3.105	−3.103	−3.095	−3.089
	2	−3.101	−3.096	−*3.110*	−3.088
	3	−3.099	−3.090	−3.088	−3.093

$p \in \bar{p}$, $q \in \bar{q}$, then for all criteria, $p_1 \geq p_0$ and $q_1 \geq q_0$, almost surely, as $T \to \infty$. However, BIC is *strongly consistent* in the sense that it will determine the true model asymptotically, whereas for AIC an overparameterised model will emerge no matter how long the available realisation. Of course, such properties are not necessarily guaranteed in finite samples, as we find below.[15]

Given the behaviour of the SACF and SPACF of our returns series, we set $\bar{p} = \bar{q} = 3$ and Table 2.4 shows the resulting AIC and BIC values. Both criteria select the orders (2,2) (an ARMA(2,2) process), although for BIC the (0,1) (an MA(1) process) is a very close second. The two estimated models are

$$x_t = 0.017 - 1.056\, x_{t-1} - 0.821\, x_{t-2} + \hat{a}_t$$
$$(0.006)\ \ (0.051)\ \ \ \ \ \ \ (0.048)$$
$$+ 1.177\, \hat{a}_{t-1} + 0.883\, \hat{a}_{t-2}, \qquad \hat{\sigma} = 5.01\%$$
$$(0.043)\ \ \ \ \ \ \ (0.039)$$

$$x_t = 0.006 + \hat{a}_t + 0.139\, \hat{a}_{t-1} \qquad \hat{\sigma} = 5.08\%$$
$$(0.002)\ \ \ \ \ \ \ (0.036)$$

Although these models appear quite different, they are, in fact, similar in two respects. The estimate of the monthly mean return implied by the ARMA(2,2) model is 0.6% (approximately 7% per annum), almost the same as that obtained directly from the MA(1) model, while the sums of the weights of the respective AR(∞) representations are very close, being 0.94 and 0.88 respectively.

There is, however, one fundamental difference between the two models: the MA(1) does not produce an acceptable fit to the returns series, for it has a $Q(12)$ value of 24.2, with a marginal significance level of 0.012. The ARMA(2,2) model, on the other hand, has an insignificant $Q(12)$ value of just 8.42.

Thus, although theoretically the BIC has advantages over the AIC, it would seem that the latter criterion nevertheless selects an identical model to that chosen by the BIC. We should also note the similarity of the complex AR and MA roots in the higher order model. These are $-0.53 \pm 0.74i$ and $-0.59 \pm 0.73i$ and this could lead to problems of parameter redundancy, with roots approximately cancelling out. From this perspective, the (2,2) model may be thought of as providing a trade-off between the parsimonious, but inadequate, (0,1) model and other, more profligately parameterised, models.

EViews Exercises

2.30 To obtain the simulated series shown in Figures 2.1, 2.2 and 2.4 open the `Chap _ 2 _ sims` page of the workfile `mills.wf1` and issue the following series of commands

```
genr a = 5*nrnd     (generates aₜ ~ NID(0,25))
smpl 1 1            (sets initial conditions for AR(1) processes)
genr x1 = a
genr x2 = a
smpl 2 100          (generates AR(1) and MA(1) processes)
genr x1 = 0.5*x1(-1) + a
genr x2 = -0.5*x2(-1) + a
genr x3 = a - 0.8*a(-1)
genr x4 = a + 0.8*a(-1)
smpl 1 2            (sets initial conditions for AR(2) processes)
genr x5 = a
genr x6 = a
genr x7 = a
genr x8 = a
smpl 3 100          (generates AR(2) and MA(2) processes)
genr x5 = 0.5*x5(-1) + 0.3*x5(-2) + a
genr x6 = x6(-1) - 0.5*x6(-2) + a
genr x7 = -0.5*x7(-1) + 0.3*x7(-2) + a
genr x8 = -0.5*x8(-1) - 0.3*x8(-2) + a
genr x9 = a + 0.5*a(-1) - 0.3*a(-2)
genr x10 = a - 0.5*a(-1) - 0.3*a(-2)
```

2.31 The *S&P 500* real returns data for Example 2.1 are contained in page `Ex _ 2 _ 1` as the variable x. The SACF and Q statistics of Table 2.1 may be obtained by opening x, clicking **View/Correlogram**, if desired changing 'Lags to include' to 12, and then clicking OK. The mean return of 3.72% may be obtained by clicking **View/Descriptive Statistics & Tests/Histogram and Stats**.

2.32 The basic data for Example 2.2 are contained in page `Ex _ 2 _ 2`: `rs` and `r20` are the short and long interest rates respectively. The spread may be computed with the command

```
genr spread = r20 - rs
```

The SACF and SPACF shown in Table 2.2 are obtained using the 'Correlogram' view of `spread` as described above.

The two forms of the AR(2) process discussed in §2.26 may be estimated with the commands

```
ls spread c spread(-1 to -2)
ls spread c ar(1 to 2)
```

It is the latter command that estimates the model in 'mean deviation' form, so providing an estimate of the mean, this being the variable `c`, and estimates of the roots, labelled 'Inverted AR roots'. In either of the 'Equation' views, clicking **View/Residual Diagnostics/Correlogram-Q-statistics** will obtain the Q-statistics for the residuals. Overfitting with AR(3) and ARMA(2,1) models may be carried out with the commands

```
ls spread c spread(-1 to -3)
ls spread c spread(-1 to -2) ma(1)
```

or

```
ls spread c ar(1 to 3)
ls spread c ar(1 to 2) ma(1)
```

2.33 The *FTA All Share* nominal return of Example 2.3 is to be found in page `Ex _ 2 _ 3` as the variable x. Note that the series runs from February 1952 as the January return is not available. The SACF and SPACF of Table 2.3 can be obtained in the usual way. Table 2.4 can be constructed by estimating all sub-models contained within the ARMA(3,3) using variants of

```
ls x c ar(1 to 3) ma(1 to 3)
```

for example

```
ls x c              (ARMA(0,0))
ls x c ar(1)        (ARMA(1,0))
ls x c ar(1 to 2)   (ARMA(2,0))
```

and so on. For each estimation, the *AIC* and *BIC* values are given by 'Akaike info criterion' and 'Schwarz criterion' respectively.

Notes

1. There are numerous introductory and intermediate level text books on time series analysis. Two that are aimed specifically at analysing economic and financial time series are Mills, *Time Series Techniques for Economists* (Cambridge University Press, 1990) and Mills and Raphael N. Markellos, *The Econometric Modelling of Financial Time Series*, 3rd edition (Cambridge University Press, 2008).
2. For more technical details on ergodicity, see Clive W.J. Granger and Paul Newbold, *Forecasting Economic Time Series*, 2nd edition (Academic Press, 1986; chapter 1) and James D. Hamilton, *Time Series Analysis* (Princeton University Press, 1994; chapter 3.2). See also Ian Domowitz and Mahmoud A. El-Gamal, 'A consistent nonparametric test of ergodicity for time series with applications', *Journal of Econometrics* 102 (2001), 365-98.
3. Herman Wold, *A Study in the Analysis of Time Series* (Almqvist and Wiksell, 1938), pages 84–9, although Wold does not refer to this theorem as a decomposition. Peter Whittle, 'Some recent contributions to the theory of stationary processes', Appendix 2 of the second edition of Wold's book, appears to be the first to refer to it as such. See also Mills, *The Foundations of Modern Time Series Analysis* (Palgrave Macmillan, 2011; chapter 7) for detailed discussion of the theorem.
4. The term 'white noise' was coined by physicists and engineers because of its resemblance, when examined in the frequency domain, to the optical spectrum of white light, which consists of very narrow lines close together: see Gwilym M. Jenkins, 'General considerations in the analysis of spectra', *Technometrics* 3 (1961), 133–66. The term 'innovation' reflects the fact that the current error a_t is, by definition, independent of all previous values of both the error and x and hence represents unforecastable 'news' becoming available at time t.
5. Autoregressions were first introduced by the famous British statistician George Udny Yule during the 1920s: see Yule, 'On a method of investigating periodicities in disturbed series, with special reference to Wolfer's sunspot numbers', *Philosophical Transactions of the Royal Society of London, Series A* 226 (1927), 267–98. For details on the historical development of these models, see Mills, *Foundations*, and Mills, *A Very British Affair: Six Britons and the Development of Time Series Analysis* (Palgrave Macmillan, 2013). Acronyms abound in time series analysis and have even prompted a journal article on them: Granger, 'Acronyms in time series analysis (ATSA)', *Journal of Time Series Analysis* 3 (1982), 103–7, although in the three decades or so since its publication many more have been suggested.
6. Moving average processes were introduced and analysed in detail by Wold, *A Study*.

7. The solution to the quadratic $\rho_1\theta^2+\theta+\rho_1= 0$ is

$$\theta = \frac{-1 \pm \sqrt{1-4\rho_1^2}}{2\rho_1}.$$

The restriction that θ be real requires that $1-4\rho_1^2 > 0$, which implies that $\rho_1^2 > 0.25$ and hence $-0.5 < \rho_1 < 0.5$.

8. As an example of this technique, consider the AR(2) process for which $\phi(B) = 1-\phi_1 B-\phi_2 B^2$. The ψ-weights are then obtained by equating coefficients in

$$(1 - \phi_1 B - \phi_2 B^2)(1+ \psi_1 B + \psi_2 B^2 + \ldots) = 1$$

or

$$1+ (\psi_1 - \phi_1)B + (\psi_2 - \phi_1\psi_1 - \phi_2)B^2 + (\psi_3 - \phi_1\psi_2 - \phi_2\psi_1)B^2 + \ldots = 1$$

For this equality to hold, the coefficients of B^j, $j \geq 0$, on each side of the equation have to be the same. Thus

$B^1: \psi_1 - \phi_1 = 0 \qquad \therefore \psi_1 = \phi_1$
$B^2: \psi_2 - \phi_1\psi_1 - \phi_2 = 0 \qquad \therefore \psi_1^2 = \phi_1^2 + \phi^2$
$B^3: \psi_3 - \phi_1\psi_2 - \phi_2\psi_1 = 0 \qquad \therefore \psi_1^3 = \phi_1^3 + 2\phi_1\phi_2$

Noting that $\psi_3 = \phi_1\psi_2 + \phi_2\psi_1$, the ψ-weights can then be derived recursively for $j \geq 2$ from $\psi_j = \phi_1\psi_{j-1} + \phi_2\psi_{j-2}$.

9. See Hamilton, *Time Series Analysis*, 17–18, for a derivation of this set of restrictions.

10. The successive sample partial autocorrelations may be estimated recursively using the updating equations proposed by James Durbin, 'The fitting of time series models', *Review of the International Statistical Institute* 28 (1960), 233–44, and which are known as the *Durbin-Levison algorithm*.

11. George E.P. Box and Jenkins, *Time Series Analysis: Forecasting and Control* (Holden Day, 1970). This is the classic reference to time series analysis: the latest edition is the fourth (Wiley, 2008), now co-authored by Gregory C. Reinsel.

12. This statistic is also known as the Ljung-Box statistic: Greta M. Ljung and Box, 'On a measure of lack of fit in time series models', *Biometrika* 65 (1978), 297–303.

13. Estimation of models with moving average errors is usually carried out by *conditional* least squares, where the initial values of the error series that are required for estimation are set to their conditional expectation of zero.

14. Hirotugu Akaike, 'A new look at the statistical model identification', *IEEE Transactions on Automatic Control* AC-19 (1974), 716–23; Gideon Schwarz, 'Estimating the dimension of a model', *Annals of Statistics* 6 (1978), 461–4.
15. See Andew Tremayne, 'Stationary linear univariate time series models', chapter 6 of Mills and Kerry Patterson (editors) *Palgrave Handbook of Econometrics, Volume 1: Theory* (Palgrave Macmillan, 2006), 215–51, for more discussion of information criteria and, indeed, for a survey of many current issues in ARMA modelling.

3
Non-stationary Time Series: Differencing and ARIMA Modelling

Non-stationarity

3.1 The class of ARMA models developed in the previous chapter relies on the assumption that the underlying process is weakly stationary, thus implying that the mean, variance and autocovariances of the process are invariant under time shifts. As we have seen, this restricts the mean and variance to be constant and requires the autocovariances to depend only on the time lag. Many economic and financial time series, however, are certainly not stationary and, in particular, have a tendency to exhibit time-changing means and/or variances.

3.2 In order to deal with such *non-stationarity*, we begin by assuming that a time series can be decomposed into a non-constant mean level plus a random error component:

$$x_t = \mu_t + \varepsilon_t \qquad (3.1)$$

A non-constant mean level μ_t in (3.1) can be modelled in a variety of ways. One potentially realistic possibility is that the mean evolves as a (non-stochastic) polynomial of order d in time. This will arise if x_t can be decomposed into a trend component, given by the polynomial, and a stochastic, stationary, but possibly autocorrelated, zero mean error component, which is always possible given Cramer's extension of Wold's decomposition theorem to non-stationary processes.[1] Thus we may have

$$x_t = \mu_t + \varepsilon_t = \sum_{j=0}^{d} \beta_j t^j + \psi(B)a_t \qquad (3.2)$$

Since

$$E(\varepsilon_t) = \psi(B)E(a_t) = 0,$$

we have

$$E(x_t) = E(\mu_t) = \sum_{j=0}^{d} \beta_j t^j$$

and, as the β_j coefficients remain constant through time, such a trend in the mean is said to be *deterministic*. Trends of this type can be removed by a simple transformation. Consider the linear trend obtained by setting $d = 1$ where, for simplicity, the error component is assumed to be a white noise sequence:

$$x_t = \beta_0 + \beta_1 t + a_t \qquad (3.3)$$

Lagging (3.3) one period and subtracting this from (3.3) itself yields

$$x_t - x_{t-1} = \beta_1 + a_t - a_{t-1} \qquad (3.4)$$

The result is a difference equation following an ARMA(1,1) process in which, since $\phi = \theta = 1$, both autoregressive and moving average roots are unity and the model is neither stationary nor invertible. If we consider the *first differences* of x_t, w_t say, then

$$w_t = x_t - x_{t-1} = (1-B)x_t = \Delta x_t$$

where $\Delta = 1 - B$ is known as the *first difference operator*. Equation (3.4) can then be written as

$$w_t = \Delta x_t = \beta_1 + \Delta a_t$$

and w_t is thus generated by a stationary, since $E(w_t) = \beta_1$ is a constant, but not invertible MA(1) process.

3.3 In general, if the trend polynomial is of order d and ε_t is characterised by the ARMA process $\phi(B)\varepsilon_t = \theta(B)a_t$, then

$$\Delta^d x_t = (1-B)^d x_t$$

obtained by first differencing x_t d times, will follow the process

$$\Delta^d x_t = \theta_0 + \frac{\Delta^d \theta(B)}{\phi(B)} a_t$$

where $\theta_0 = d!\beta_d$. Thus the MA part of the process generating $\Delta^d x_t$ will also contain the factor Δ^d and will therefore have d roots of unity. Note also that the variance of x_t will be the same as the variance of ε_t and so will be constant for all t. Figure 3.1 shows plots of generated data for both linear and quadratic trend models. Because the variance of the error component, here assumed to be white noise

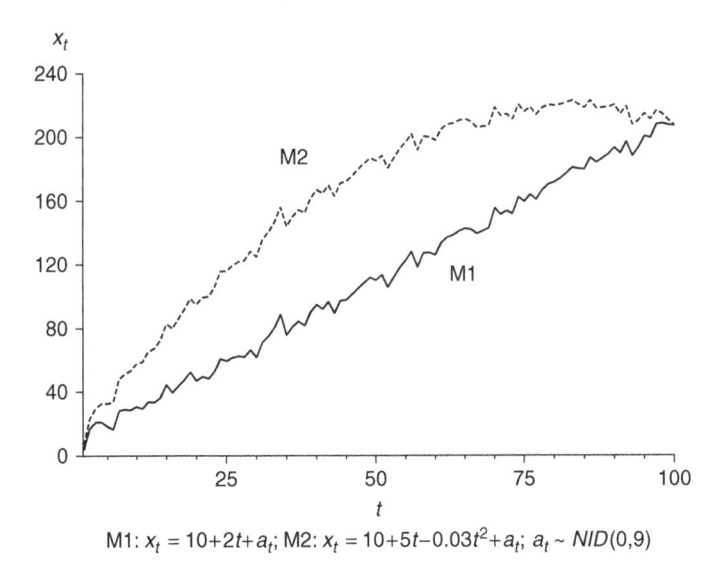

M1: $x_t = 10+2t+a_t$; M2: $x_t = 10+5t-0.03t^2+a_t$; $a_t \sim NID(0,9)$

Figure 3.1 Linear and quadratic trends

and distributed as $NID(0,9)$, is constant and independent of the level, the variabilities of the two series are bounded about their expected values, and the trend components are clearly observed in the plots.

3.4 An alternative way of generating a non-stationary mean level is to employ ARMA models whose autoregressive parameters do not satisfy stationarity conditions. For example, consider the AR(1) process

$$x_t = \phi x_{t-1} + a_t \tag{3.5}$$

where $\phi > 1$. If the process is assumed to have started at time $t = 0$, the difference equation (3.5) has the solution

$$x_t = x_0 \phi^t + \sum_{i=0}^{t} \phi^i a_{t-i} \tag{3.6}$$

The 'complementary function' $x_0 \phi^t$ can be regarded as the **conditional expectation** of x_t at time $t = 0$ and is clearly an increasing function of t. The conditional expectation of x_t at subsequent times $t = 1, 2, ...$ depends on the sequence of random shocks $a_1, a_2, ...$ and hence, since this conditional expectation may be regarded as the trend of x_t, the trend changes **stochastically**.

The variance of x_t is given by

$$V(x_t) = \sigma^2 \frac{\phi^{2t} - 1}{\phi^2 - 1}$$

which is also an increasing function of time and becomes infinite as $t \to \infty$.[2] In general, x_t will have a trend in both mean and variance, and such processes are said to be **explosive**. A plot of generated data from the process (3.5) with $\phi = 1.05$ and $a_t \sim NID(0,9)$, and having starting value $x_0 = 10$, is shown in Figure 3.2. We see that, after a short 'induction period', the series essentially follows an exponential curve with the generating a_ts playing almost no further part. The same behaviour would be observed if additional autoregressive and moving average terms were added to the model, as long as the stationarity conditions are violated.

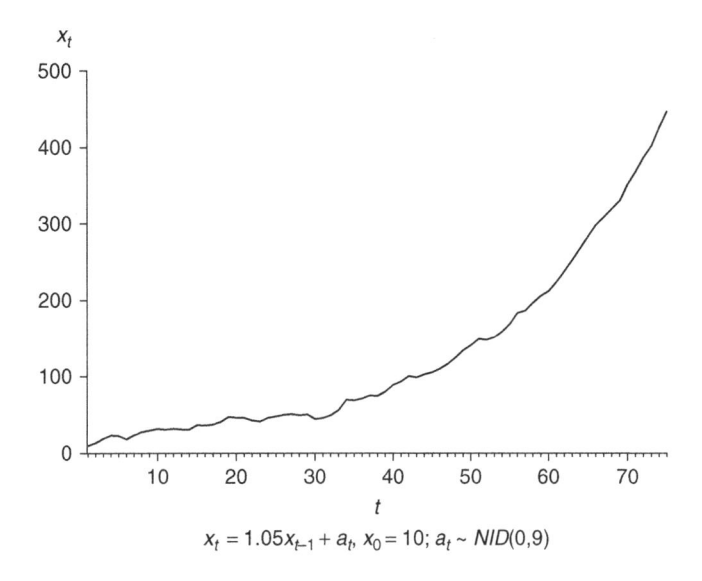

Figure 3.2 Explosive AR(1) model

$x_t = 1.05x_{t-1} + a_t, x_0 = 10; a_t \sim NID(0,9)$

ARIMA processes

3.5 As we can see from (3.6), the solution of (3.5) is explosive if $\phi > 1$ but stationary if $\phi < 1$. The case $\phi = 1$ provides a process that is neatly balanced between the two. If x_t is generated by the model

$$x_t = x_{t-1} + a_t \tag{3.7}$$

then a_t is said to follow a **random walk**.[3] If we allow a constant, θ_0, to be included, so that

$$x_t = x_{t-1} + \theta_0 + a_t \tag{3.8}$$

then x_t will follow a **random walk with drift**. If the process starts at $t = 0$, then

$$x_t = x_0 + t\theta_0 + \sum_{i=0}^{t} a_{t-i},$$

so that

$$\mu_t = E(x_t) = x_0 + t\theta_0$$

$$\gamma_{0,t} = V(x_t) = t\sigma^2$$

and

$$\gamma_{k,t} = Cov(x_t, x_{t-k}) = (t-k)\sigma^2 \quad k \geq 0$$

are all functions of t and hence are time varying.

3.6 The correlation between x_t and x_{t-k} is then given by

$$\rho_{k,t} = \frac{\gamma_{k,t}}{\sqrt{\gamma_{0,t}\gamma_{0,t-k}}} = \frac{t-k}{\sqrt{t(t-k)}} = \sqrt{\frac{t-k}{t}}$$

If t is large compared to k, all the $\rho_{k,t}$ will be approximately unity. The sequence of x_t values will therefore be very smooth, but will also be non-stationary since both the mean and variance of x_t will change with t. Figure 3.3 shows generated plots of the random walks (3.7) and (3.8) with $x_0 = 10$ and $a_t \sim NID(0,9)$. In part (a) of the figure the drift parameter, θ_0, is set to zero while in part (b) we have set $\theta_0 = 2$. The two plots differ considerably, but neither show any affinity with the initial value x_0: indeed, the expected length of time for a random walk to pass again through an arbitrary value is infinite.

3.7 The random walk is an example of a class of non-stationary models known as **integrated processes**. Equation (3.8) can be written as

$$\Delta x_t = \theta_0 + a_t$$

and so first differencing x_t leads to a stationary model, in this case the white noise process a_t. Generally, a series may need first differencing d times to attain stationarity, and the series so obtained may itself be autocorrelated.

(a) $x_t = x_{t-1} + a_t$, $x_0 = 10$; $a_t \sim NID(0,9)$

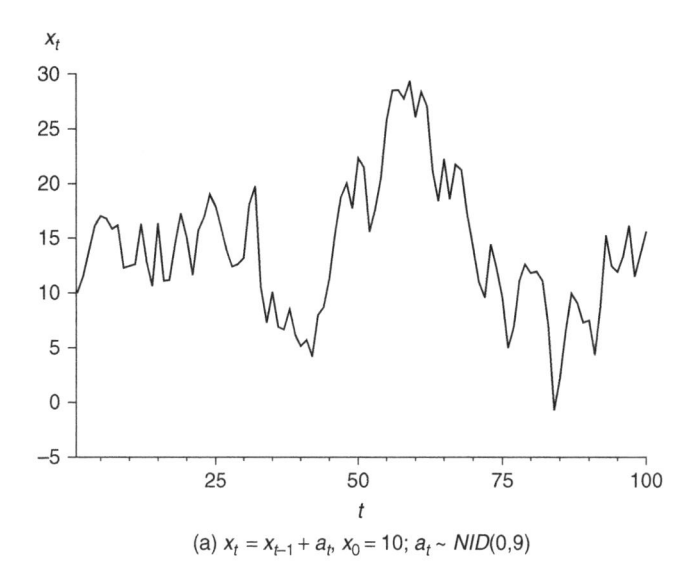

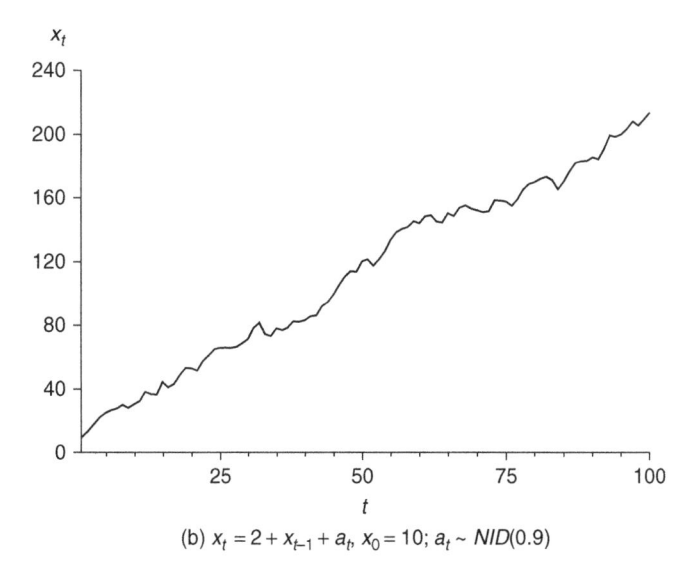

(b) $x_t = 2 + x_{t-1} + a_t$, $x_0 = 10$; $a_t \sim NID(0.9)$

Figure 3.3 Random walks

If this autocorrelation is modelled by an ARMA(p,q) process, then the model for the original series is of the form

$$\phi(B)\Delta^d x_t = \theta_0 + \theta(B)a_t \qquad (3.9)$$

which is said to be an **autoregressive-integrated-moving average** process of orders p, d and q, or ARIMA(p,d,q), and x_t is said to be integrated of order d, denoted $I(d)$.[4]

3.8 It will usually be the case that the order of integration d or, equivalently, the degree of differencing, will be 0, 1 or, very occasionally, 2. Again it will be the case that the autocorrelations of an ARIMA process will be close to 1 for all non-large k. For example, consider the (stationary) ARMA(1,1) process

$$x_t - \phi x_{t-1} = a_t - \theta a_{t-1}$$

whose ACF has been shown to be (§2.24)

$$\rho_1 = \frac{(1-\phi\theta)(\phi-\theta)}{1+\theta^2 - 2\phi\theta} \qquad \rho_k = \phi\rho_{k-1} \quad \text{for} \quad k > 1$$

As $\phi \to 1$, the ARIMA(0,1,1) process

$$\Delta x_t = a_t - \theta a_{t-1}$$

results, and all the ρ_k tend to unity.

3.9 A number of points concerning the ARIMA class of models are of importance. Consider again (3.9), with $\theta_0 = 0$ for simplicity:

$$\phi(B)\Delta^d x_t = \theta(B)a_t$$

This process can equivalently be defined by the two equations

$$\phi(B)w_t = \theta(B)a_t$$

and

$$w_t = \Delta^d x_t \qquad (3.10)$$

so that, as we have noted above, the model corresponds to assuming that $\Delta^d x_t$ can be represented by a stationary and invertible ARMA process. Alternatively, for $d \geq 1$, (3.10) can be inverted to give

$$x_t = S^d w_t \qquad (3.11)$$

where S is the infinite summation, or *integral*, operator defined by

$$S = \left(1 + B + B^2 + \ldots\right) = (1-B)^{-1} = \Delta^{-1}$$

Equation (3.11) implies that x_t can be obtained by summing, or 'integrating', the stationary series w_t d times: hence the term integrated process.

3.10 This type of non-stationary behaviour is often referred to as **homogenous non-stationarity**, and it is important to discuss why this form of non-stationarity is felt to be useful when describing the behaviour of many economic and financial time series. Consider again the first-order autoregressive process (3.2). A basic characteristic of the AR(1) model is that, for both $|\phi| < 1$ and $\phi > 1$, the 'local' behaviour of a series generated from the model is heavily dependent upon the level of x_t. In the former case local behaviour will always be dominated by an affinity to the mean, while in the latter the series will eventually increase rapidly with t. For many economic and financial series, however, local behaviour appears to be roughly independent of level, and this is what we mean by homogenous non-stationarity.

3.11 If we want to use ARMA models for which the behaviour of the process is indeed independent of its level, then the autoregressive polynomial $\phi(B)$ must be chosen so that

$$\phi(B)(x_t + c) = \phi(B)x_t$$

where c is any constant. Thus

$$\phi(B)c = 0$$

implying that $\phi(1) = 0$, so that $\phi(B)$ must be able to be factorised as

$$\phi(B) = \phi_1(B)(1-B) = \phi_1(B)\Delta,$$

in which case the class of processes that need to be considered will be of the form

$$\phi_1(B)w_t = \theta(B)a_t$$

where $w_t = \Delta x_t$. Since the requirement of homogenous non-stationarity precludes w_t increasing explosively, either $\phi_1(B)$ is a stationary operator, or $\phi_1(B) = \phi_2(B)(1-B)$, so that $\phi_2(B)w_t^* = \theta(B)a_t$, where $w_t^* = \Delta^2 x_t$. Since this argument can be used recursively, it follows that for time series that are homogenously non-stationary, the autoregressive lag polynomial must be of the form $\phi(B)\Delta^d$, where $\phi(B)$ is a stationary polynomial. Figure 3.4 plots generated data from the model $\Delta^2 x_t = a_t$, where $a_t \sim NID(0,9)$ and $x_0 = x_1 = 10$, and such a series is seen to display random movements in both level and slope.

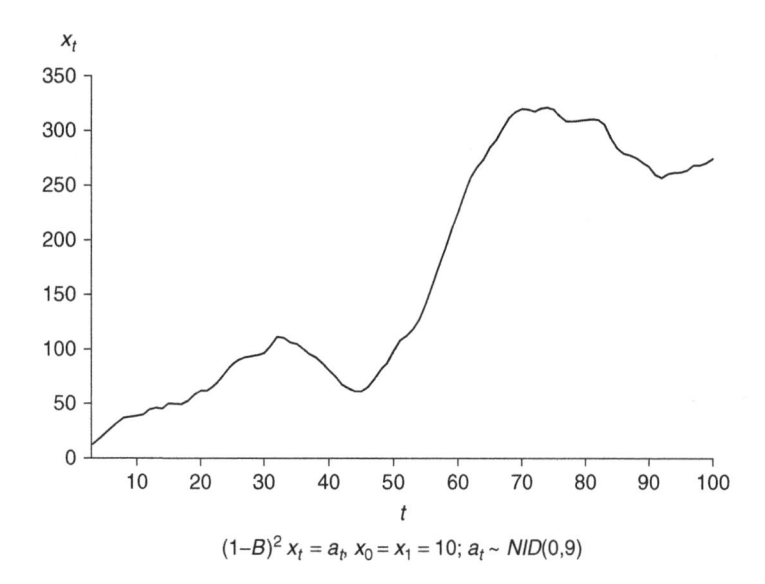

Figure 3.4 'Second difference' model

Non-stationary Time Series 51

3.12 In general, if a constant is included in the model for dth differences, then a deterministic polynomial trend of degree d is automatically allowed for. Equivalently, if θ_0 is allowed to be non-zero, then

$$E(w_t) = E(\Delta^d x_t) = \mu_w = \theta_0/(1-\phi_1-\phi_2-\ldots-\phi_p)$$

is non-zero, so that an alternative way of expressing (3.9) is as

$$\phi(B)\tilde{w}_t = \theta(B)a_t$$

where $\tilde{w}_t = w_t - \mu_w$.

Figure 3.5 plots generated data for $\Delta^2 x_t = 2 + a_t$, where again $a_t \sim NID(0,9)$ and $x_0 = x_1 = 10$. The inclusion of the deterministic quadratic trend has a dramatic effect on the evolution of the series, with the non-stationary 'noise' being completely swamped after a few periods.

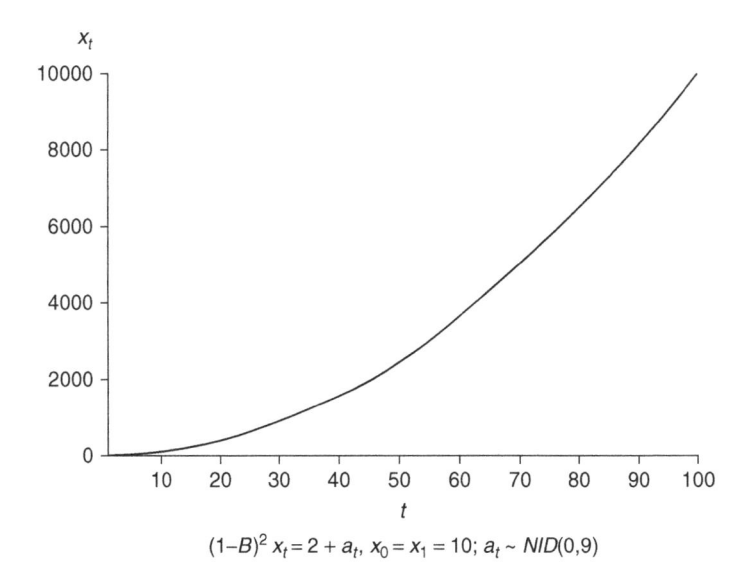

$(1-B)^2 x_t = 2 + a_t$, $x_0 = x_1 = 10$; $a_t \sim NID(0,9)$

Figure 3.5 'Second difference with drift' model

Model (3.9) therefore allows both stochastic and deterministic trends to be modelled. When $\theta_0 = 0$ a stochastic trend is incorporated, while if $\theta_0 \neq 0$ the model may be interpreted as representing a deterministic trend (a polynomial in time of order d) buried in non-stationary noise, which will thus be autocorrelated. The models presented in §§3.2–3.3 could be described as deterministic trends buried in *stationary* noise, since they can be written as

$$\phi(B)\Delta^d x_t = \phi(1)\beta_d d! + \Delta^d \theta(B) a_t$$

Here the stationary nature of the noise in the level of x_t is manifested in d roots of the moving average lag polynomial being unity.

ARIMA modelling

3.13 Once the order of differencing d has been established then, since $w_t = \Delta^2 x_t$ is by definition stationary, the ARMA model building techniques discussed in §§2.27–2.29 may be applied to the suitably differenced series. Establishing the correct order of differencing is by no means straightforward, however, and is discussed in detail in §§4.4–4.7. We content ourselves here with a sequence of examples illustrating the modelling of ARIMA processes when d has already been chosen: the suitability of these choices will be examined through examples in Chapter 4.

EXAMPLE 3.1 Modelling the UK spread as an integrated process

In Example 2.2 we modelled the spread of UK interest rates as a stationary, indeed AR(2), process. Here we consider modelling the spread assuming that it is an $I(1)$ process, so that we examine the behaviour of the SACF and SPACF of $w_t = \Delta x_t$. Table 3.1 provides these estimates up to $k = 12$ and suggests that, as both functions cut off at $k = 1$, either an AR(1) or an MA(1) process is identified. Estimation of the former obtains

$$w_t = -0.0013 + 0.208\ w_{t-1} + \hat{a}_t, \qquad \hat{\sigma} = 0.405$$
$$\quad\ \ (0.0147)\ \ (0.036)$$

Table 3.1 SACF and SPACF of the first difference of the UK spread

k	r_k	$se(r_k)$	$\hat{\phi}_{kk}$	$se(\hat{\phi}_{kk})$
1	0.208	0.036	0.208	0.036
2	−0.027	0.038	−0.017	0.036
3	−0.019	0.038	−0.023	0.036
4	0.018	0.038	0.028	0.036
5	0.035	0.038	0.027	0.036
6	0.003	0.038	−0.011	0.036
7	−0.016	0.038	−0.014	0.036
8	−0.072	0.039	−0.067	0.036
9	−0.082	0.039	−0.058	0.036
10	−0.037	0.039	−0.009	0.036
11	−0.009	0.039	−0.002	0.036
12	−0.020	0.039	0.024	0.036

The residuals are effectively white noise, as they yield a portmanteau statistic of $Q(12) = 8.62$, and the mean of w_t is seen to be insignificantly different from zero. The spread can thus be modelled as an ARIMA(1,1,0) process without drift. In fact, fitting an ARIMA(0,1,1) process obtains almost identical estimates, with θ estimated to be 0.204 and $\hat{\sigma} = 0.405$.

The implication of this model is that the spread evolves as a driftless random walk with AR(1) innovations. Being non-stationary, the spread therefore has no equilibrium level to return to and thus 'wanders widely' but without any drift up or, indeed, down. All innovations consequently have *permanent* effects, in direct contrast to the AR(2) model of Example 2.2, in which the spread is stationary about an equilibrium level so that, since the series always reverts back to this level, all innovations can have only *temporary* effects. A method of distinguishing between these alternative models is introduced in Chapter 4.

EXAMPLE 3.2 Modelling the dollar/sterling ($/£) exchange rate

Figure 3.6 plots monthly observations of both the level and first differences of the $/£ exchange rate from January 1973 to December 2014, a total of 504 observations. The levels exhibit the wandering

Levels

Differences

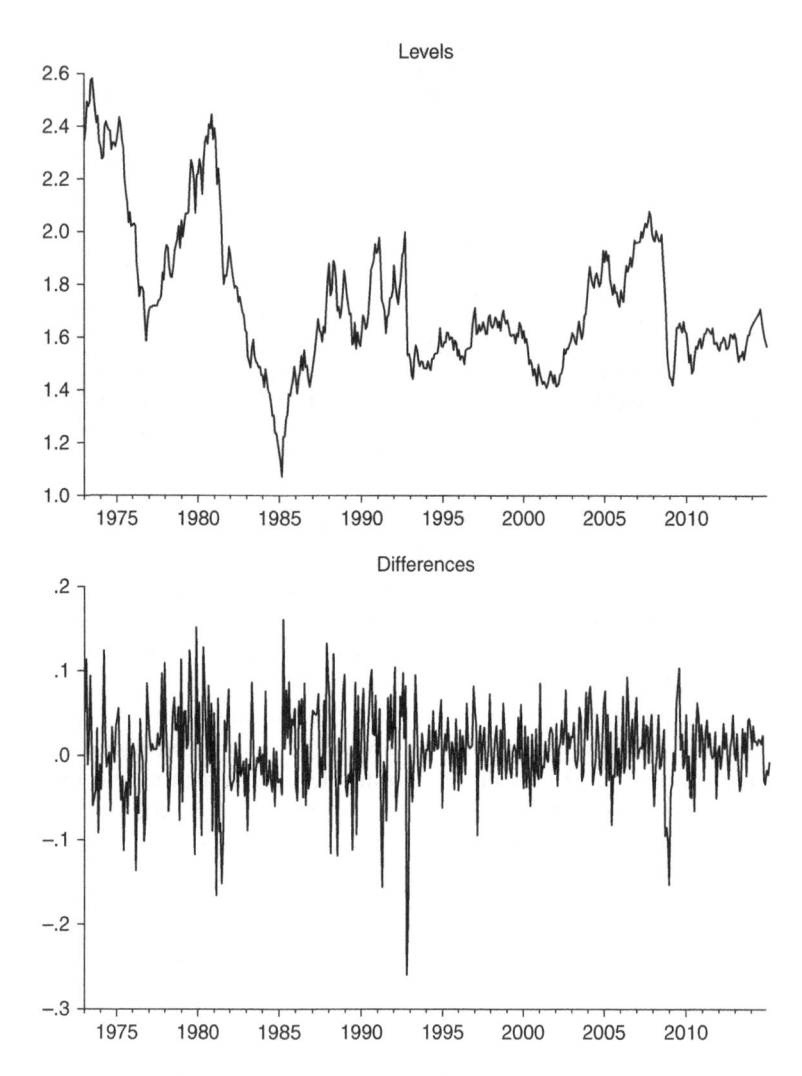

Figure 3.6 $/£ exchange rate (January 1973–December 2014)

movement of a driftless random walk: the SACF has $r_1 = 0.978$, $r_2 = 0.952$, $r_3 = 0.922$, $r_6 = 0.832$ and $r_{12} = 0.669$ and thus displays the slow, almost linear, decline typical of an $I(1)$ process (this is discussed further in §4.2).

Non-stationary Time Series 55

The differences are stationary about zero and appear to show no discernable pattern: indeed, they are very close to being a white noise process, the only significant sample autocorrelation being $r_1 = 0.128$. Although the parameter estimates are significant on fitting either an AR(1) or MA(1) process, the R^2 statistic associated with each model is just 0.016, which, of course, is approximately equal to r_1^2.

EViews Exercises

3.14 The linear and quadratic trends of Figure 3.1 can be generated by opening page `Chap _ 3 _ sims` of `mills.wf1` and issuing the commands

```
genr a = 3*nrnd
genr t = @trend + 1
genr x1 = 10 + 2*t + a
genr x2 = 10 + 5*t - 0.03*(t^2) + a
```

The explosive AR(1) model of Figure 3.2 is obtained with

```
smpl 1 1
genr x3 = 10 + a
smpl 2 100
genr x3 = 1.05*x3(-1) + a
```

Similarly, the random walks of Figure 3.3 are obtained with the commands

```
smpl 1 1
genr x4 = 10 + a
genr x5 = 12 + a
smpl 2 100
genr x4 = x4(-1) + a
genr x5 = 2 + x5(-1) + a
```

The 'second difference' models of Figures 3.4 and 3.5 are obtained with

```
smpl 1 1
genr x6 = 10
genr x7 = 10
smpl 2 2
genr x6 = 2*x6(-1) - 10 + a
genr x7 = 2*x7(-1) - 8 + a
smpl 3 100
genr x6 = 2*x6(-1) - x6(-2) + a
genr x7 = 2 + 2*x7(-1) - x7(-2) + a
```

3.15 In Example 3.1 the SACF and SPACF of Table 3.1 are obtained by opening the page `Ex _ 3 _ 1`, opening the variable `spread` and obtaining the correlogram in the usual way, except in this case 'First difference' is selected in the Correlogram Specification.

Estimates of the AR(1) and MA(1) models for the first difference of the spread are obtained with the commands

```
ls d(spread) c d(spread(-1))
ls d(spread) c ma(1)
```

respectively.

3.16 The $/£ exchange rate is found in page `Ex _ 3 _ 2` as the variable `dollar`. Figure 3.6 shows this variable and its first difference, `d(dollar)`. The SACFs reported in Example 3.2 are obtained in the usual way by opening `dollar` and obtaining the correlogram for the levels and first differences respectively.

Notes

1. Harold Cramer, 'On some classes of non-stationary processes', *Proceedings of the 4th Berkeley Symposium on Mathematical Statistics and Probability* (University of California Press, 1961), 57–78.

2. The variance of x_t is

$$V(x_t) = E(x_t^2) = E(a_t + \phi a_{t-1} + \phi^2 a_{t-2} + \ldots + \phi^{2t-1} a_1)^2$$
$$= \sigma^2(1 + \phi^2 + \phi^4 + \ldots + \phi^{2t-1})$$
$$= \sigma^2 \frac{1 - \phi^{2t}}{1 - \phi^2} = \sigma^2 \frac{\phi^{2t} - 1}{\phi^2 - 1}$$

on using the white noise assumptions and the standard result that

$$1 + z + z^2 + \ldots + z^{t-1} = (1 - z^t)/(1 - z)$$

with $z = \phi^2$.

3. The term random (or drunkard's) walk was coined by Karl Pearson in correspondence with Lord Rayleigh in the journal *Nature* in 1905. Although first employed by Pearson to describe a mosquito infestation in a forest, the model was subsequently, and memorably, used to describe the optimal 'search strategy' for finding a drunk who had been left in the middle of a field at the dead of night! The solution is to start exactly where the drunk had been placed, as that point is an unbiased estimate of the drunk's future position, and then walk in a randomly selected straight line, since he will presumably stagger along in an unpredictable and random fashion.

Pearson's metaphor was, of course, in terms of *spatial* displacement, but the time series analogy should be clear. Random walks were, in fact, first formally introduced in continuous time by Louis Bachelier in his 1900 doctoral dissertation *Theorie de Speculation* in order to describe the unpredictable evolution of stock prices. They were independently discovered by Albert Einstein in 1905 and have since played a fundamental role in mathematics and physics as models of, for example, waiting times, limiting diffusion processes, and first-passage problems.

4. This terminology was introduced in Box and Jenkins, *Time Series Analysis*.

4
Unit Roots and Related Topics

Determining the order of integration of a time series

4.1 As we have shown in §§3.5–3.12, the order of integration, d, is a crucial determinant of the properties that a time series exhibits. If we restrict ourselves to the most common values of 0 and 1 for d, so that x_t is either $I(0)$ or $I(1)$, then it is useful to bring together the properties of such processes.

If x_t is $I(0)$, which we will denote $x_t \sim I(0)$ even though such a notation has been used previously to denote the distributional characteristics of a series, then, assuming for convenience that x_t has zero mean,

(i) the variance of x_t is finite and does not depend on t;
(ii) the innovation a_t has only a temporary effect on the value of x_t;
(iii) the expected length of times between crossings of $x = 0$ is finite, so that x_t fluctuates around its mean of zero;
(iv) the autocorrelations, ρ_k, decrease steadily in magnitude for large enough k, so that their sum is finite.

If $x_t \sim I(1)$ with $x_0 = 0$, then

(i) the variance of x_t goes to infinity as t goes to infinity;
(ii) an innovation a_t has a permanent effect on the value of x_t because x_t is the sum of all previous innovations: see, for example, equation (2.16);
(iii) the expected time between crossings of $x = 0$ is infinite;
(iv) the autocorrelations $\rho_k \to 1$ for all k as t goes to infinity.

Unit Roots and Related Topics 59

4.2 The fact that a time series is non-stationary is often self-evident from a plot of the series. Determining the actual form of non-stationarity, however, is not so easy from just a visual inspection and an examination of the SACFs for various differences may be required.
To see why this may be so, recall from §2.15 that a stationary AR(p) process requires that all roots g_i in

$$\phi(B) = (1 - g_1 B)(1 - g_2 B) \ldots (1 - g_p B)$$

be such that $|g_i| < 1$. Now suppose that one of them, say g_1, approaches 1, so that $g_1 = 1 - \delta$, where δ is a small positive number. The autocorrelations

$$\rho_k = A_1 g_1^k + A_2 g_2^k + \ldots + A_p g_p^k \cong A_1 g_1^k$$

will then be dominated by $A_1 g_1^k$, since all other terms will go to zero more rapidly. Moreover, because g_1 is close to 1, the exponential decay $A_1 g_1^k$ will be slow and almost linear:

$$A_1 g_1^k = A_1(1 - \delta)^k = A(1 - \delta k + \delta^2 k^2 - \ldots) \cong A_1(1 - \delta k)$$

Hence failure of the SACF to die down quickly is an indication of non-stationarity, its behaviour tending to be rather that of a slow, linear decline. If the original series x_t is found to be non-stationary, the first difference Δx_t is then analysed. If Δx_t is still non-stationary, the next difference $\Delta^2 x_t$ is analysed, the procedure being repeated until a stationary difference is found, although in practice d will not exceed 2.

4.3 Sole reliance on the SACF can sometimes lead to problems of *over-differencing*. Although further differences of a stationary series will themselves be stationary, over-differencing can lead to serious difficulties. Consider the stationary MA(1) process $x_t = (1 - \theta B)a_t$. The first difference of this is

$$\Delta x_t = (1 - B)(1 - \theta B)a_t$$
$$= (1 - (1 + \theta)B + \theta B^2)a_t$$
$$= (1 - \theta_1 B - \theta_2 B^2)a$$

We now have a more complicated model containing two parameters rather than one and, moreover, one of the roots of the $\theta(B)$ polynomial will be unity since $\theta_1 + \theta_2 = 1$. The model is therefore not invertible, so that the AR(∞) representation does not exist and attempts to estimate this model will almost surely run into difficulties.

Testing for a unit root

4.4 Given the importance of choosing the correct order of differencing, it is clear that we need to have available a formal testing procedure to determine d. To introduce the issues involved in developing such a procedure, we begin by considering the simplest case, that of the zero mean AR(1) process:

$$x_t = \phi x_{t-1} + a_t \qquad t = 1, 2, \ldots, T \tag{4.1}$$

where $a_t \sim WN(0, \sigma^2)$ and $x_0 = 0$. The OLS estimator of ϕ is given by

$$\hat{\phi}_T = \frac{\sum_{t=1}^{T} x_{t-1} x_t}{\sum_{t=1}^{T} x_t^2}$$

A conventional way of testing the null hypothesis $\phi = 1$ is to construct the t-statistic

$$t_\phi = \frac{\hat{\phi}_T - 1}{\hat{\sigma}_{\hat{\phi}_T}} = \frac{\hat{\phi}_T - 1}{\left(s_T^2 / \sum_{t=1}^{T} x_{t-1}^2\right)^{1/2}} \tag{4.2}$$

where

$$\hat{\sigma}_{\hat{\phi}_T} = \left(s_T^2 / \sum_{t=1}^{T} x_{t-1}^2\right)^{1/2}$$

is the usual OLS standard error for $\hat{\phi}_T$ and s_T^2 is the OLS estimator of σ^2:

$$s_T^2 = \sum_{t=1}^{T} (x_t - \hat{\phi}_T x_{t-1})^2 \Big/ (T-1)$$

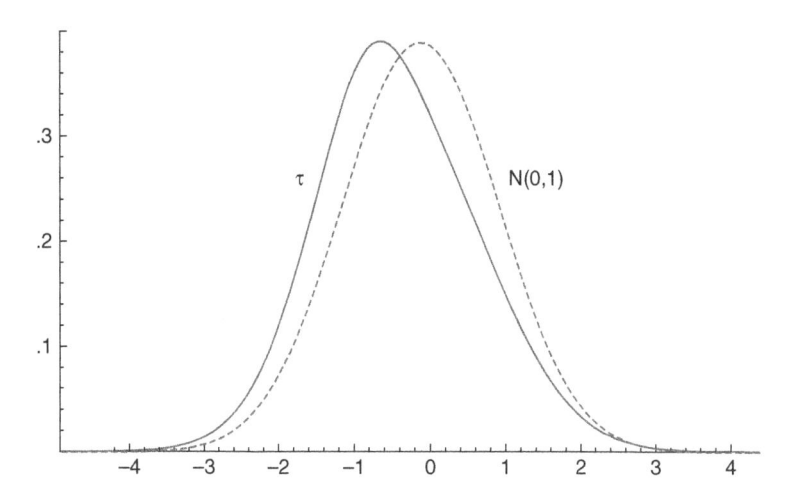

Figure 4.1 Limiting Distribution of τ

Unfortunately, the distribution of t_ϕ does not have a limiting normal distribution when $\phi = 1$. Rather, its distribution is shown in Figure 4.1, where it is called the τ-distribution in recognition of its non-normality. The test statistic (4.2) is renamed τ rather than t_ϕ and is often known as the **Dickey-Fuller** test, as indeed is the distribution.[1]

Figure 4.1 shows that the limiting distribution of τ is approximately standard normal but shifted to the left by roughly 0.3: the large T 5%, 2.5% and 1% critical values for τ are −1.95, −2.23 and −2.58, rather than the standard normal critical values of −1.65, −1.96 and −2.33.

4.5 This case has the merit of being simple, but is not particularly realistic, for it implies that the alternative to a driftless random walk is a stationary AR(1) process about a *zero* mean, which would rule out most economic and financial time series, which can typically only take on positive values. A more sensible alternative would thus be for the AR(1) process to fluctuate about a non-zero mean, so that we have the model

$$x_t = \theta_0 + \phi x_{t-1} + a_t \qquad t = 1, 2, \ldots, T \tag{4.3}$$

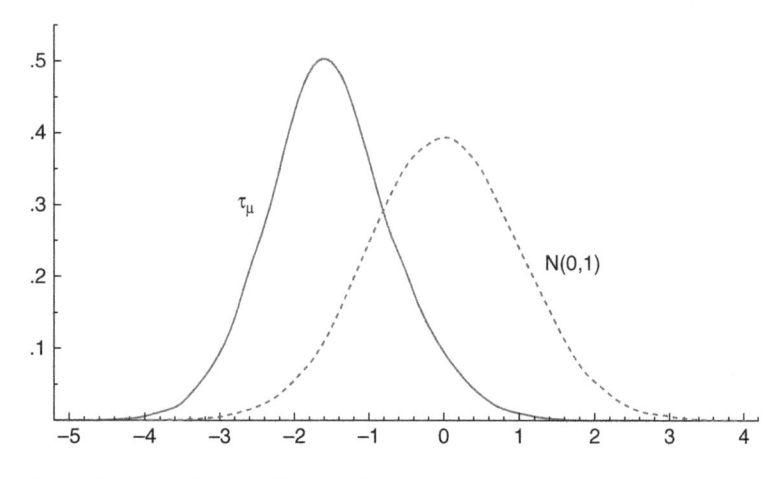

Figure 4.2 Limiting distribution of τ_μ

in which the unit root null is parameterised as $\theta_0 = 0$, $\phi = 1$. The presence of an intercept alters the distribution of the test statistic, which is now denoted τ_μ to emphasise that a non-zero mean is allowed for in the regression (4.3).[2] Figure 4.2 presents the simulated distribution of τ_μ. With a non-zero mean, the distribution under the unit root null deviates further from the standard normal than when the mean is zero (compare with Figure 4.1), with the large T 5%, 2.5% and 1% critical values now being −2.86, −3.12 and −3.43.

4.6 A further generalisation is to allow the innovations to be auto correlated. Suppose that x_t is generated by the AR(p) process

$$(1 - \phi_1 B - \phi_2 B^2 - \ldots - \phi_p B^p)x_t = \theta_0 + a_t$$

or

$$x_t = \theta_0 + \sum_{i=1}^{p} \phi_i x_{t-i} + a_t \qquad (4.4)$$

A more convenient representation is obtained by defining

$$\phi = \sum_{i=1}^{p} \phi_i$$

$$\delta_i = -\sum_{j=i+1}^{p-1} \phi_j \qquad i = 1, 2, \ldots, p-1$$

so that (4.4) can be written, with $k = p-1$,

$$x_t = \theta_0 + \phi x_{t-1} + \sum_{i=1}^{k} \delta_i \Delta x_{t-i} + a_t \qquad (4.5)$$

The null of one unit root is thus $\phi = \sum_{i=1}^{p} \phi_i = 1$.[3] OLS provides consistent estimators of the coefficients of (4.5) and a test of $\phi = 1$ can be constructed as

$$\tau_\mu = \frac{\hat{\phi}_T - 1}{se(\hat{\phi}_T)}$$

where $se(\hat{\phi}_T)$ is the OLS standard error attached to the estimate $\hat{\phi}_T$ (recall §4.4). This statistic is also denoted τ_μ because it has the *same* limiting distribution as the statistic obtained from the AR(1) model (4.3), although it is often referred to as the **augmented Dickey-Fuller** (ADF) test. In a similar vein, (4.5) is known as the **augmented Dickey-Fuller regression**.

4.7 The above analysis has implicitly assumed that the AR order p is known, so that we are certain that x_t is generated by a pth order autoregression. If the generating process is an ARMA(p,q), then the τ_μ statistic obtained from estimating the model

$$x_t = \theta_0 + \phi x_{t-1} + \sum_{i=1}^{k} \delta_i \Delta x_{t-i} + a_t - \sum_{j=1}^{q} \theta_j a_{t-j}$$

has the same limiting distribution as that calculated from (4.5). The problem here, of course, is that p and q are assumed known, and this is unlikely to be the case in practice. When p and q are unknown, the test statistic obtained from (4.5) can still be used if k, the number of lags of Δx_t introduced as regressors, increases with the sample size. With typical economic and financial data, setting k at $[T^{0.25}]$ should work well in practice, where $[\cdot]$ denotes the operation of taking the integer part of the argument: for example, for $T = 50$, $T^{0.25} = 2.659$, so that k is set at 2; for $T = 500$, $T^{0.25} = 4.729$ and hence $k = 4$. This adjustment is necessary because, as the sample size increases, the effects of the correlation structure of the residuals on the shape of the distribution of τ_μ become more precise. A more accurate setting of k may be determined by using, for example, information criteria given a maximum value of k of $[T^{0.25}]$.

EXAMPLE 4.1 Unit root tests on the spread and the $/£ exchange rate

Examples 2.2 and 3.1 examined two models for the UK interest rate spread, a stationary AR(2) process and an $I(1)$ process without drift. We are now in a position to discriminate between the two through the application of a unit root test. The fitted AR(2) model

$$x_t = 0.036 + 1.192\ x_{t-1} - 0.224\ x_{t-2} + a_t$$
$$(0.017)\ \ (0.036)\ \ \ \ \ \ \ \ (0.036)$$

can equivalently be written as

$$x_t = 0.036 + 0.969\ x_{t-1} + 0.224\ \Delta x_{t-1} + a_t$$
$$(0.017)\ \ (0.008)\ \ \ \ \ \ \ \ (0.036)$$

so that $\tau_\mu = (0.969 - 1)/0.008 = -3.93$, which is significant at the 1% level, this critical value being -3.44. Note that the τ_μ statistic can be obtained directly as the t-ratio on x_{t-1} from rewriting the model again as

$$\Delta x_t = 0.036 - 0.031\ x_{t-1} + 0.224\ \Delta x_{t-1} + a_t$$
$$(0.017)\ \ (0.008)\ \ \ \ \ \ \ \ (0.036)$$

We may therefore conclude that the spread does not contain a unit root and that the appropriate specification is the stationary AR(2) model in which there are temporary, albeit highly persistent, deviations away from an equilibrium level of 1.128%.

A similar approach to testing for a unit root in the $/£ exchange rate, the presence of which was assumed in Example 3.2, leads to the estimated equation

$$\Delta x_t = 0.039 - 0.024\ x_{t-1} + 0.136\ \Delta x_{t-1} + a_t$$
$$(0.014)\ \ (0.008)\ \ \ \ \ \ \ \ (0.044)$$

Unit Roots and Related Topics   65

Here we have $\tau_\mu = -2.93$ and, since the 5% critical value is -2.87, this is (just) significant, although it is not significant at the 1% level (the marginal significance level is 0.043). Thus there is some doubt as to whether the appropriate model for the \$/£ exchange rate is indeed a (possibly autocorrelated) random walk or whether it is stationary around an 'equilibrium' rate, estimated here to be 1.672. This latter model would have the implication that any deviation from this level would only be temporary and foreign exchange traders would then have a 'one-way' bet in that such deviations must eventually be reversed, which seems highly unlikely in such a competitive and efficient market.[4] A resolution of this apparent 'paradox' will be provided in Example 5.1.

Trend versus difference stationarity

4.8 In the unit root testing strategy outlined above, the implicit null hypothesis is that the series is generated as a driftless random walk with, possibly, serially correlated innovations. In popular terminology, x_t is said to be ***difference stationary*** (DS),[5]

$$\Delta x_t = \varepsilon_t \tag{4.6}$$

where $\varepsilon_t = \theta(B)a_t$, while the alternative is that x_t is *stationary* in levels. While the null of a driftless random walk is appropriate for many financial time series such as interest rates and exchange rates, other economic series often do contain a drift, so that the relevant null becomes

$$\Delta x_t = \theta + \varepsilon_t \tag{4.7}$$

In this case, a plausible alternative is that x_t is generated by a linear trend buried in stationary noise (see §3.12), now termed ***trend stationarity*** (TS)

$$x_t = \beta_0 + \beta_1 t + \varepsilon_t \tag{4.8}$$

Unfortunately, the τ_μ statistic obtained from (4.5) is incapable of distinguishing a stationary process around a linear trend (model

(4.8)) from a process with a unit root and drift (model (4.7)). Indeed, rejection of a null hypothesis of a unit root is unlikely using this statistic if the series is stationary around a linear trend and becomes impossible as the sample size increases.[6]

4.9 A test of (4.7) against (4.8) is, however, straightforward to carry out by using an extension of the testing methodology discussed above: the ADF regression (4.5) is simply extended by the inclusion of the time trend t as an additional regressor,

$$x_t = \beta_0 + \beta_1 t + \phi x_{t-1} + \sum_{i=1}^{k} \delta_i \Delta x_{t-i} + a_t \quad (4.9)$$

and the statistic

$$\tau_\tau = \frac{\hat{\phi}_T - 1}{se(\hat{\phi}_T)}$$

is calculated. This 't-statistic' is denoted τ_τ to distinguish it from τ_μ because it has a different limiting distribution, which is shown in Figure 4.3. The large T 5%, 2.5% and 1% critical values are now −3.41, −3.66 and −3.96.

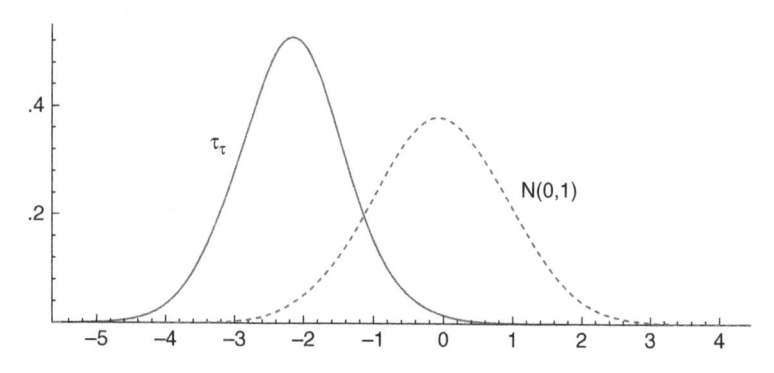

Figure 4.3 Limiting distribution of τ_τ

EXAMPLE 4.2 Are UK equity prices trend or difference stationary?

In Example 2.3 we modelled the returns of the UK *FTA All Share* index as an ARMA process. In fact, the returns were defined as Δx_t, where x_t is the **logarithm** of the index. Thus, by analysing returns, we are implicitly assuming that the logarithm of the index is $I(1)$ and that x_t is DS. Figure 4.4 plots these logarithms, which are seen to have a pronounced tendency to drift upwards, albeit with some major 'wanderings' about trend, most notably over the last decade and a half or so of the sample period. We may thus investigate whether this DS representation is appropriate or whether a TS model would be preferable.

Setting the lag augmentation at $k = 3$ led to the ADF regression

$$\Delta x_t = \underset{(0.021)}{0.051} + \underset{(0.00004)}{0.00007}\, t - \underset{(0.006)}{0.012}\, x_{t-1} + \sum_{i=1}^{3} \hat{\delta}_i \Delta x_{t-i} + \hat{a}_t$$

This yields the test statistic $\tau_\tau = -2.15$. Since the 10% critical value is -3.13, there is thus no evidence against the hypothesis that the logarithm of the index is DS, confirming the use of returns in modelling

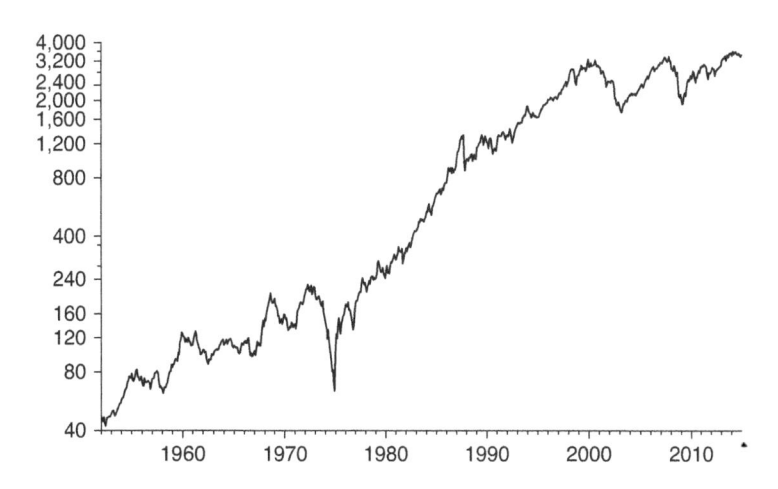

Figure 4.4 FTA All Share index on a logarithmic scale (January 1952–December 2014)

this series. If the logarithms of the index had been TS, this would have implied that they evolved as autocorrelated deviations about a linear trend, again providing traders with a one-way bet whenever the index got too far away from this trend. Even a cursory examination of Figure 4.4 shows that such a representation is clearly false.

Testing for more than one unit root

4.10 The above development of unit root tests has been predicated on the assumption that x_t contains *at most* one unit root, so that it is at most $I(1)$. If the null hypothesis of a unit root is not rejected, then it may be necessary to test whether the series contains a second unit root, in other words whether it is $I(2)$ and thus needs differencing twice to induce stationarity.

EXAMPLE 4.3 Do UK interest rates contain two unit roots?

Figure 4.5 shows plots of the UK short and long interest rates from which the spread, analysed in Example 4.1, is calculated. To test for the presence of one unit root in each of the interest rates we estimate the regressions

$$\Delta RS_t = 0.063 - 0.010\ RS_{t-1} + 0.303\ \Delta RS_{t-1}$$
$$(0.032)\quad(0.004)\qquad\quad(0.035)$$

$$\Delta R20_t = 0.025 - 0.004\ R20_{t-1} + 0.303\ \Delta R20_{t-1} - 0.122\ \Delta R20_{t-2}$$
$$(0.026)\quad(0.003)\qquad\quad(0.003)\qquad\quad(0.003)$$

where RS_t and $R20_t$ are the short and long rates, respectively. The τ_μ statistics are thus −2.28 and −1.15, so confirming that both interest rates are $I(1)$.

To test for a second unit root, the following Dickey-Fuller regressions were run for the first differences of the interest rates:

$$\Delta^2 RS_t = -\ 0.702\ \Delta RS_{t-1}$$
$$(0.035)$$

$$\Delta^2 R20_t = -\ 0.823\ \Delta R20_{t-1} + 0.125\ \Delta^2 R20_{t-1}$$
$$(0.044)\qquad\qquad(0.036)$$

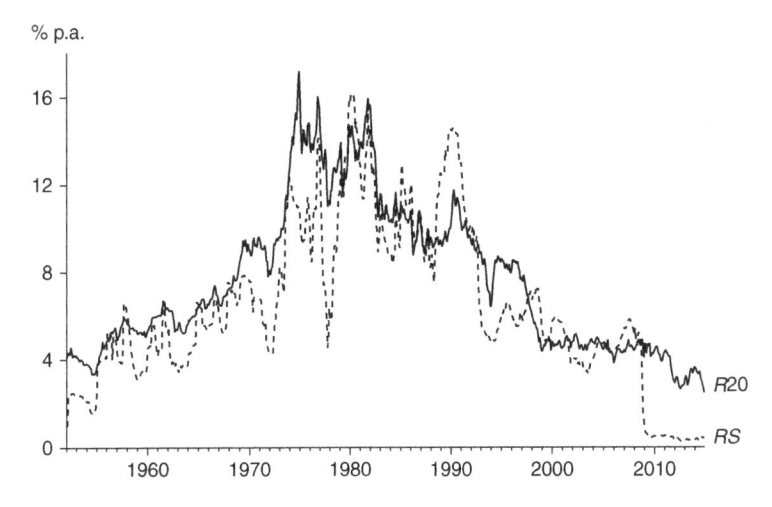

Figure 4.5 UK interest rates (January 1952–December 2014)

The τ_μ statistics are computed here to be -20.2 and -18.8, thus conclusively rejecting the hypothesis of two unit roots in both series.

EViews Exercises

4.11 The unit root tests in Example 4.1 can be computed automatically. To test for a unit root in the spread, open page Ex_3_1 of workfile mills.wf1 and open the variable spread. Selecting *View/Unit Root test…* and clicking OK will obtain the ADF-statistic and the differenced form of the ADF regression. The lag length in the ADF regression can either be chosen by the user or selected automatically by a variety of information criteria with the maximum lag length set at $[\min(T/3,12) \times (T/100)^{0.25}]$. A similar procedure for the dollar variable in page Ex_3_2 will obtain the ADF test statistic and regression for the $/£ exchange rate.

4.12 The values of the *FTA All Share* index are found in page Ex_4_2 as the variable index. The logarithms of the index may be obtained with the command genr p = log(index). To test whether these logarithms are TS or DS open p and conduct a unit root test but this time checking 'Trend and intercept' in the 'Include in test equation' box and choosing a lag length of 3. The lag length could also be selected automatically but the inference will remain unchanged: the ADF statistic is insignificant and the series is DS.

4.13 The unit root tests for the short and long interest rates of Example 4.3 may be obtained in a similar fashion to those for the spread and exchange rate using the data in page Ex _ 2 _ 2, although to test for a second unit root, select '1st difference' and 'none' in the appropriate boxes.

Notes

1. The seminal article on what has become a vast topic, and which gives the distribution and test their eponymous names, is David A. Dickey and Wayne A. Fuller, 'Distribution of the estimators for autoregressive time series with a unit root', *Journal of the American Statistical Association* 74 (1979), 427–31. The statistical theory underlying the distribution is too advanced to be considered here but see, for example, Patterson, *A Primer for Unit Root Testing* (Palgrave Macmillan, 2010) and, at a rather more technical level, his *Unit Root Tests in Time Series, Volume 1: Key Concepts and Problems* (Palgrave Macmillan, 2011). As will be seen from these texts, there is now a considerable number of unit root tests, differing in their size and power properties. Nevertheless, the original Dickey-Fuller tests remain popular and widely used.
2. Strictly, τ_μ tests $\phi = 1$ *conditional* upon $\theta_0 = 0$, so that the model under the null is the driftless random walk $x_t = x_{t-1} + a_t$. The joint hypothesis $\theta_0 = 0$, $\phi = 1$ may be tested by constructing a standard F-test, although clearly the statistic, typically denoted Φ, will not follow the $F(2,T-2)$ distribution. For large samples, the 5% and 1% critical values of the appropriate distribution are 4.59 and 6.53, rather than the 2.99 and 4.60 critical values of the F-distribution: see Dickey and Fuller, 'Likelihood ratio statistics for autoregressive time series with a unit root', *Econometrica* 49 (1981), 1057–72.
3. This generalisation is most clearly seen when $p = 2$, so that

$$x_t = \theta_0 + \phi_1 x_{t-1} + \phi_2 x_{t-2} + a_t$$

This can be written as

$$\begin{aligned} x_t &= \theta_0 + \phi_1 x_{t-1} + \phi_2 x_{t-1} - \phi_2 x_{t-1} + \phi_2 x_{t-2} + a_t \\ &= \theta_0 + (\phi_1 + \phi_2) x_{t-1} - \phi_2 \Delta x_{t-1} + a_t \\ &= \theta_0 + \phi x_{t-1} + \delta_1 \Delta x_{t-1} + a_t \end{aligned}$$

which is (4.5) with $k = 1$.

4. The mean deviation form of the implied stationary AR(2) model for the $/£ rate is estimated to be

$$x_t = 1.672 + 1.113(x_{t-1} - 1.672) - 0.133(x_{t-2} - 1.672) + a_t$$

which has two real roots of 0.97 and 0.14.
5. This terminology was introduced in Charles R. Nelson and Charles I. Plosser, 'Trends and random walks in macroeconomic time series', *Journal of Monetary Economics* 10 (1982), 139–62.
6. These results were provided by Pierre Perron, 'Trends and random walks in macroeconomic time series: further evidence from a new approach', *Journal of Economic Dynamics and Control* 12 (1988), 297–332.

5
Modelling Volatility using GARCH Processes

Volatility

5.1 Following the initial work on portfolio theory in the 1950s, volatility has become an extremely important concept in finance, appearing regularly in models of, for example, asset pricing and risk management. Much of the interest in volatility has to do with it not being directly observable, and several alternative measures have been developed to approximate it empirically. The most common measure of volatility has been the unconditional standard deviation of historical returns. The use of this measure, however, is severely limited by it not necessarily being an appropriate representation of financial risk and by the fact that returns tend not to be independent and identically distributed, so making the standard deviation a potentially poor estimate of underlying volatility.

An alternative approach to measuring volatility is to embed it within a formal stochastic model for the time series of returns. A simple way to do this is to allow the variance (or the conditional variance) of the process generating the returns series x_t to change either continuously or at certain discrete points in time. Although a stationary process must have a constant variance, certain conditional variances can change, so that although the unconditional variance $V(x_t)$ may be constant for all t, the conditional variance $V(x_t|x_{t-1}, x_{t-2},...)$, which depends on the realisation of x_t, is able to alter from period to period.

Modelling Volatility using GARCH Processes 73

5.2 A simple way to develop a stochastic model having time varying conditional variances is to suppose that x_t is generated by the **product process**

$$x_t = \mu + \sigma_t U_t \tag{5.1}$$

where U_t is a **standardised process**, so that $E(U_t) = 0$ and $V(U_t) = E(U_t^2) = 1$ for all t, and σ_t is a sequence of positive random variables such that

$$V(x_t \,|\sigma_t) = E((x_t - \mu)^2 \,|\sigma_t) = \sigma_t^2 E(U_t^2) = \sigma_t^2$$

σ_t^2 is thus the **conditional variance** and σ_t the **conditional standard deviation** of x_t.

Typically $U_t = (x_t - \mu)/\sigma_t$ is assumed to be normal and independent of σ_t: we will further assume that it is strict white noise, so that $E(U_t U_{t-k}) = 0$ for $k \neq 0$. These assumptions imply that x_t has mean μ, variance

$$E(x_t - \mu)^2 = E(\sigma_t^2 U_t^2) = E(\sigma_t^2)E(U_t^2) = E(\sigma_t^2)$$

and autocovariances

$$E(x_t - \mu)(x_{t-k} - \mu) = E(\sigma_t \sigma_{t-k} U_t U_{t-k}) = E(\sigma_t \sigma_{t-k})E(U_t U_{t-k}) = 0$$

and is thus white noise. However, note that both the squared and absolute deviations, $S_t = (x_t - \mu)^2$ and $M_t = |x_t - \mu|$, can be autocorrelated. For example,

$$\begin{aligned}Cov(S_t, S_{t-k}) &= E(S_t - E(S_t))(S_{t-k} - E(S_t)) = E(S_t S_{t-k}) - (E(S_t))^2 \\ &= E(\sigma_t^2 \sigma_{t-k}^2)E(U_t^2 U_{t-k}^2) - (E(\sigma_t^2))^2 \\ &= E(\sigma_t^2 \sigma_{t-k}^2) - (E(\sigma_t^2))^2\end{aligned}$$

so that

$$E(S_t^2) = E(\sigma_t^4) - (E(\sigma_t^2))^2$$

and the kth autocorrelation of S_t is

$$\rho_{k,S} = \frac{E(\sigma_t^2 \sigma_{t-k}^2) - (E(\sigma_t^2))^2}{E(\sigma_t^4) - (E(\sigma_t^2))^2}$$

This autocorrelation will only be zero if σ_t^2 is constant, in which case x_t can be written as $x_t = \mu + a_t$, where $a_t = \sigma U_t$ has zero mean and constant variance σ, which is just another way of defining a_t, and hence x_t, to be white noise.

ARCH processes

5.3 So far we have said nothing about how the conditional variances σ_t^2 might be generated. We now consider the case where they are a function of past values of x_t:

$$\sigma_t^2 = f(x_{t-1}, x_{t-2}, \ldots)$$

A simple example is

$$\sigma_t^2 = f(x_{t-1}) = \alpha_0 + \alpha_1 (x_{t-1} - \mu)^2 \tag{5.2}$$

where α_0 and α_1 are both positive. With $U_t \sim NID(0,1)$ and independent of σ_t, $x_t = \mu + \sigma_t U_t$ is then white noise and conditionally normal,

$$x_t \mid x_{t-1}, x_{t-2}, \ldots \sim NID(\mu, \sigma_t^2)$$

so that

$$V(x_t \mid x_{t-1}) = \alpha_0 + \alpha_1 (x_{t-1} - \mu)^2$$

If $0 < \alpha_1 < 1$ then the unconditional variance is $V(x_t) = \alpha_0/(1 - \alpha_1)$ and x_t is weakly stationary. It may be shown that the fourth moment of x_t is finite if $3\alpha_1^2 < 1$ and, if so, the kurtosis of x_t is given by $3(1-\alpha_1^2)/(1 - 3\alpha_1^2)$. Since this must exceed 3, the unconditional distribution of x_t is fatter tailed than the normal. If this moment condition is not satisfied, then the variance of x_t^2 will be infinite and x_t^2 will not be weakly stationary.

5.4 This model is known as the *first-order autoregressive conditional heteroskedastic* (ARCH(1)) process.[1] ARCH processes have proved to be an extremely popular class of non-linear models for economic and financial time series. A more convenient notation is to define $\varepsilon_t = x_t - \mu = U_t\sigma_t$, so that the ARCH(1) model can be written as

$$\varepsilon_t | x_{t-1}, x_{t-2}, \ldots \sim NID(0, \sigma_t^2)$$

$$\sigma_t^2 = \alpha_0 + \alpha_1 \varepsilon_{t-1}^2$$

On defining $v_t = \varepsilon_t^2 - \sigma_t^2$, the model can also be written as

$$\varepsilon_t^2 = \alpha_0 + \alpha_1 \varepsilon_{t-1}^2 + v_t$$

Since $E(v_t | x_{t-1}, x_{t-2}, \ldots) = 0$, the model corresponds directly to an AR(1) model for the squared innovations ε_t^2. However, as $v_t = \sigma_t^2 (U_t^2 - 1)$, the errors are obviously heteroskedastic.

5.5 A natural extension is to the ARCH(*q*) process, where (5.2) is replaced by

$$\sigma_t^2 = f(x_{t-1}, x_{t-2}, \ldots, x_{t-q}) = \alpha_0 + \sum_{i=1}^{q} \alpha_i (x_{t-i} - \mu)^2$$

where $\alpha_0 \geq 0$ and $\alpha_i > 0$, $1 \leq i \leq q$. The process will be weakly stationary if all the roots of the characteristic equation associated with the ARCH parameters are less than unity. This implies that $\sum_{i=1}^{q} \alpha_i < 1$, in which case the unconditional variance is $V(x_t) = \alpha_0/(1 - \sum_{i=1}^{q} \alpha_i)$. In terms of ε_t and σ_t^2, the conditional variance function is

$$\sigma_t^2 = \alpha_0 + \sum_{i=1}^{q} \alpha_i \varepsilon_{t-i}^2$$

or, equivalently, on defining $\alpha(B) = \alpha_1 + \alpha_2 B + \ldots + \alpha_q B^{q-1}$,

$$\varepsilon_t^2 = \alpha_0 + \alpha(B)\varepsilon_{t-1}^2 + v_t$$

5.6 A practical difficulty with ARCH models is that, with q large, unconstrained estimation will often lead to violation of the non-negativity constraints on the α_is that are needed to ensure that the conditional variance σ_t^2 is always positive. In many early applications of the model a rather arbitrary declining lag structure was imposed on the α_is to ensure that these constraints were met. To obtain more flexibility, we consider a further extension, to the *generalised ARCH* (GARCH) process.[2] The GARCH(p,q) process has the conditional variance function

$$\sigma_t^2 = \alpha_0 + \sum_{i=1}^q \alpha_i \varepsilon_{t-i}^2 + \sum_{i=1}^p \beta_i \sigma_{t-i}^2$$
$$= \alpha_0 + \alpha(B)\varepsilon_{t-1}^2 + \beta(B)\sigma_{t-1}^2$$

where $p > 0$ and $\beta_i \geq 0$, $i \leq 1 \leq p$. For the conditional variance of the GARCH(p,q) process to be well defined, all the coefficients in the corresponding ARCH(∞) model $\sigma_t^2 = \theta_0 + \theta(B)\varepsilon_t^2$ must be positive. Provided that $\alpha(B)$ and $\beta(B)$ have no common roots and that the roots of $1-\beta(B)$ are all less than unity, this positivity constraint will be satisfied if and only if all the coefficients in $\theta(B) = \alpha(B)/(1-\beta(B))$ are non-negative. For the GARCH(1,1) process,

$$\sigma_t^2 = \alpha_0 + \alpha_1 \varepsilon_{t-1}^2 + \beta_1 \sigma_{t-1}^2$$

a model that has proved extremely popular for modelling financial time series, these conditions require that all three parameters are non-negative.

The equivalent form of the GARCH(p,q) process is

$$\varepsilon_t^2 = \alpha_0 + (\alpha(B) + \beta(B))\varepsilon_{t-1}^2 + v_t - \beta(B)v_{t-1} \qquad (5.3)$$

so that $\varepsilon_t^2 \sim$ ARMA(m,p), where $m = \max(p,q)$. This process will be weakly stationary if and only if the roots of $1-\alpha(B)-\beta(B)$ are all less than unity, so that $\alpha(1) + \beta(1) < 1$.

5.7 If $\alpha(1) + \beta(1) = 1$ in (5.3) then $1-\alpha(B)-\beta(B)$ contains a unit root and we say that the model is *integrated GARCH*, or IGARCH. It is often the case that $\alpha(1) + \beta(1)$ is very close to unity for financial time series and, if this condition holds, a shock to the conditional variance is persistent in the sense that it remains important for all future observations.

5.8 Although we have assumed that the distribution of ε_t is conditionally normal, this is not essential. For example, the distribution could be Student's-t with unknown degrees of freedom v that may be estimated from the data: for $v > 2$ such a distribution is leptokurtic and hence has thicker tails than the normal. Whatever the assumed error distribution, estimation will require non-linear iterative techniques and maximum likelihood estimation is available in many econometric packages.

5.9 The analysis has also proceeded on the further assumption that $\varepsilon_t = x_t - \mu$ is serially uncorrelated. A natural extension is to allow x_t to follow an ARMA(P,Q) process, so that the combined ARMA(P,Q)–GARCH(p,q) model becomes

$$\Phi(B)(x_t - \mu) = \Theta(B)\varepsilon_t$$

$$\sigma_t^2 = \alpha_0 + \alpha(B)\varepsilon_{t-1}^2 + \beta(B)\sigma_{t-1}^2$$

Testing for the presence of ARCH errors

5.10 Let us suppose that an ARMA model for x_t has been estimated, from which the residuals e_t have been obtained. The presence of ARCH may lead to serious model misspecification if it is ignored: as with all forms of heteroskedasticity, analysis assuming its absence will result in inappropriate parameter standard errors, and these will typically be too small. For example, ignoring ARCH will lead to the identification of ARMA models that tend to be overparameterised.

5.11 Methods for testing whether ARCH is present are therefore essential, particularly as estimation incorporating it requires complicated iterative techniques. Equation (5.3) has shown that if ε_t is GARCH(p,q) then ε_t^2 is ARMA(m,p), where $m = \max(p,q)$, and standard ARMA theory follows through in this case. This implies that the squared residuals e_t^2 can then be used to identify m and p, and therefore q, in a similar fashion to the way the residuals themselves are used in conventional

ARMA modelling. For example, the sample autocorrelations of e_t^2 have asymptotic variance T^{-1} and portmanteau statistics calculated from them are asymptotically χ^2 if the ε_t^2 are independent.

5.12 Formal tests are also available. A test of the null hypothesis that ε_t has a constant conditional variance against the alternative that the conditional variance is given by an ARCH(q) process, which is a test of $\alpha_1 = ... = \alpha_q = 0$ conditional upon $\beta_1 = ... = \beta_p = 0$, may be based on the Lagrange Multiplier (LM) principle. The test procedure is to run a regression of e_t^2 on $e_{t-1}^2,..., e_{t-q}^2$ and to test the statistic $T \cdot R^2$ as a χ_q^2 variate, where R^2 is the squared multiple correlation coefficient of the regression. An asymptotically equivalent form of the test, which may have better small sample properties, is to compute the standard F test from the regression.[3] The intuition behind this test is clear. If the data are indeed homoskedastic, then the variance is constant and variations in e_t^2 will be purely random. If ARCH effects are present, however, such variations will be predicted by lagged values of the squared residuals.

Of course, if the residuals themselves contain some remaining autocorrelation or, perhaps, some other form of non-linearity, then it is quite likely that this test for ARCH will reject, since these misspecifications may induce autocorrelation in the squared residuals. We cannot simply assume that ARCH effects are necessarily present when the ARCH test rejects.

5.13 When the alternative is a GARCH(p,q) process, some complications arise. In fact, a general test of $p > 0$, $q > 0$ against a white noise null is not feasible, nor is a test of GARCH($p + r_1$, $q + r_2$) errors, where $r_1 > 0$ and $r_2 > 0$, when the null is GARCH(p,q). Furthermore, under this null, the LM test for GARCH(p,r) and ARCH($p + r$) alternatives coincide. What can be tested is the null of an ARCH(p) process against a GARCH(p,q) alternative.[4]

5.14 Several modifications to the standard GARCH model result from allowing the relationship between σ_t^2 and ε_t to be more flexible than the quadratic relationship that has so far been assumed. To simplify the exposition, we shall concentrate on variants of the GARCH(1,1) process

$$\sigma_t^2 = \alpha_0 + \alpha_1 \varepsilon_{t-1}^2 + \beta_1 \sigma_{t-1}^2 = \alpha_0 + \alpha_1 \sigma_{t-1}^2 U_{t-1}^2 + \beta_1 \sigma_{t-1}^2 \tag{5.4}$$

An early alternative was to model conditional standard deviations rather than variances:

$$\sigma_t = \alpha_0 + \alpha_1|\varepsilon_{t-1}| + \beta_1\sigma_{t-1} = \alpha_0 + \alpha_1\sigma_{t-1}|U_{t-1}| + \beta_1\sigma_{t-1} \quad (5.5)$$

This makes the conditional variance the square of a weighted average of absolute shocks, rather than the weighted average of squared shocks. Consequently, large shocks have a smaller effect on the conditional variance than in the standard GARCH model.[5] Rather than concentrating on the variance or standard deviation, a more flexible and general class of *power* GARCH models can be obtained by estimating an additional parameter:[6]

$$\sigma_t^\gamma = \alpha_0 + \alpha_1\left|\varepsilon_{t-1}\right|^\gamma + \beta_1\sigma_{t-1}^\gamma$$

5.15 An asymmetric response to shocks is made explicit in the *exponential GARCH* (EGARCH) model[7]

$$\log(\sigma_t^2) = \alpha_0 + \alpha_1 g(\varepsilon_{t-1}/\sigma_{t-1}) + \beta_1 \log(\sigma_{t-1}^2) \quad (5.6)$$

where

$$g(\varepsilon_{t-1}/\sigma_{t-1}) = \theta_1 \varepsilon_{t-1}/\sigma_{t-1} + \left(\left|\varepsilon_{t-1}/\sigma_{t-1}\right| - E\left|\varepsilon_{t-1}/\sigma_{t-1}\right|\right)$$

The 'news impact curve', $g(\cdot)$, relates conditional volatility, here given by $\log(\sigma_t^2)$, to 'news', ε_{t-1}. It embodies an asymmetric response since $\partial g/\partial \varepsilon_{t-1} = 1 + \theta_1$ when $\varepsilon_{t-1} > 0$ and $\partial g/\partial \varepsilon_{t-1} = 1 - \theta_1$ when $\varepsilon_{t-1} < 0$ (note that volatility will be at a minimum when there is no news, $\varepsilon_{t-1} = 0$). This asymmetry is potentially useful as it allows volatility to respond more rapidly to falls in a market than to corresponding rises, which is an important stylised fact for many financial assets and is known as the leverage effect. The EGARCH model also has the advantage that no parameter restrictions are necessary in order to ensure that the variance is positive. It is easy to show that $g(\varepsilon_{t-1}/\sigma_{t-1})$ is strict white noise with zero mean and constant variance, so that $\log(\sigma_t^2)$ is an ARMA(1,1) process and will be stationary if $\beta_1 < 1$.

5.16 A model which nests (5.4), (5.5) and (5.6) is the *non-linear ARCH* (NARCH) model, a general form of which is

$$\sigma_t^\gamma = \alpha_0 + \alpha_1 g^\gamma(\varepsilon_{t-1}) + \beta_1 \sigma_{t-1}^\gamma$$

while an alternative is the threshold ARCH process

$$\sigma_t^\gamma = \alpha_0 + \alpha_1 h^{(\gamma)}(\varepsilon_{t-1}) + \beta_1 \sigma_{t-1}^\gamma$$

where

$$h^{(\gamma)}(\varepsilon_{t-1}) = \theta_1 |\varepsilon_{t-1}|^\gamma 1(\varepsilon_{t-1} > 0) + |\varepsilon_{t-1}|^\gamma 1(\varepsilon_{t-1} \leq 0)$$

$1(\cdot)$ being the indicator function which takes the value 1 if the argument is satisfied and 0 when it is not. If $\gamma = 1$, we have the *threshold ARCH* (TARCH) model, while for $\gamma = 2$ we have the *GJR* model, which allows a quadratic response of volatility to news but with different coefficients for good and bad news, although it maintains the assertion that the minimum volatility will result when there is no news.[8]

5.17 An alternative formalisation of the GARCH(1,1) model (5.4) is to define $\alpha_1 = \varpi(1 - \alpha_1 - \beta_1)$, where ϖ is the unconditional variance, or long-run volatility, to which the process reverts to:

$$\sigma_t^2 = \varpi + \alpha_1(\varepsilon_{t-1}^2 - \varpi) + \beta_1(\sigma_{t-1}^2 - \varpi)$$

This formalisation may be extended to allow reversion to a varying level defined by q_t:

$$\sigma_t^2 = q_t + \alpha_1(\varepsilon_{t-1}^2 - q_{t-1}) + \beta_1(\sigma_{t-1}^2 - q_{t-1})$$

$$q_t = \varpi + \xi(q_{t-1} - \varpi) + \zeta(\varepsilon_{t-1}^2 - \sigma_{t-1}^2)$$

Here q_t is the permanent component of volatility which converges to ϖ through powers of ξ, while $\sigma_t^2 - q_t$ is the transitory component, converging to 0 via powers of $\alpha_1 + \beta_1$. This *component GARCH* model can also be combined with TARCH to allow asymmetries in both the permanent and transitory parts: this *asymmetric component*

GARCH model automatically introduces asymmetry into the transitory equation.[9]

5.18 There are many other variants to the basic GARCH model but these typically require specialised software to estimate and cannot be treated here.

Example 5.1   GARCH models for the $/£ exchange rate

Table 5.1 presents the results of fitting various AR(1)-GARCH(p,q) models to the first differences of the $/£ exchange rate, Δx_t. The choice of an AR(1) model for the conditional mean equation is based on our findings from Examples 3.2 and 4.1. Assuming homoskedastic (GARCH(0,0)) errors produces the estimates in the first column of Table 5.1. The ARCH(1) statistic, the LM test for first-order ARCH, shows that there is strong evidence of conditional heteroskedasticity.

A GARCH(1,1) conditional variance is fitted in the second column, using the estimation technique of quasi-maximum likelihood (QML). Both GARCH parameters are significant, and the LM test for any neglected ARCH is insignificant. The GARCH parameters sum to just under unity, suggesting that shocks to the conditional variance are very persistent. Note that the AR coefficient in the mean equation is now insignificant: its previous significance in earlier examples is thus seen to be a consequence of incorrectly assuming that the errors were homoskedastic rather than generated as a GARCH process. The estimated 'pure' GARCH(1,1) model is shown in the third column: the exchange rate is thus generated as a driftless random walk with

Table 5.1 $/£ exchange rate: QML estimates

	GARCH(0,0)	GARCH(1,1)	GARCH(1,1)
$\hat{\phi}_1$	0.129 (2.91)	0.070 (1.32)	–
$\hat{\alpha}_0$	–	7.38 (2.10)	8.30 (2.13)
$\hat{\alpha}_1$	–	0.133 (3.86)	0.144 (3.87)
$\hat{\beta}_0$	–	0.842 (25.8)	0.828 (23.4)
$\hat{\alpha}_1 + \hat{\beta}_1$	–	0.975	0.972
ARCH(1)	30.2 [.00]	0.5 [.48]	0.7 [.42]
Log-L	796.7	829.3	829.7

Figures in (...) are t-statistics; Figures in [...] are marginal significance levels.
Log-L is the log-likelihood. Estimates of α_0 are scaled by ϕ^5.

GARCH(1,1) errors (note that overfitting using GARCH(1,2) and GARCH(2,1) models proved unsuccessful).

Finally, if the lagged level of the exchange rate is added to the mean equation then this will provide a test of a unit root under GARCH(1,1) errors: doing so yields a coefficient estimate of −0.0005 with a t-statistic of just −0.41. The paradox found in Example 4.1 thus disappears: once the error is correctly specified as a GARCH process, there is no longer any evidence against the hypothesis that the exchange rate is a random walk.

The conditional standard deviations from this model are shown in Figure 5.1 along with the differences Δx_t. Large values of $\hat{\sigma}_t$ are seen to match up with periods of high volatility in the exchange rate, most notably around the UK's departure from the Exchange Rate Mechanism (ERM) in September 1992 and during the financial crisis of 2008–2009, in which the \$/£ rate dropped by over a quarter

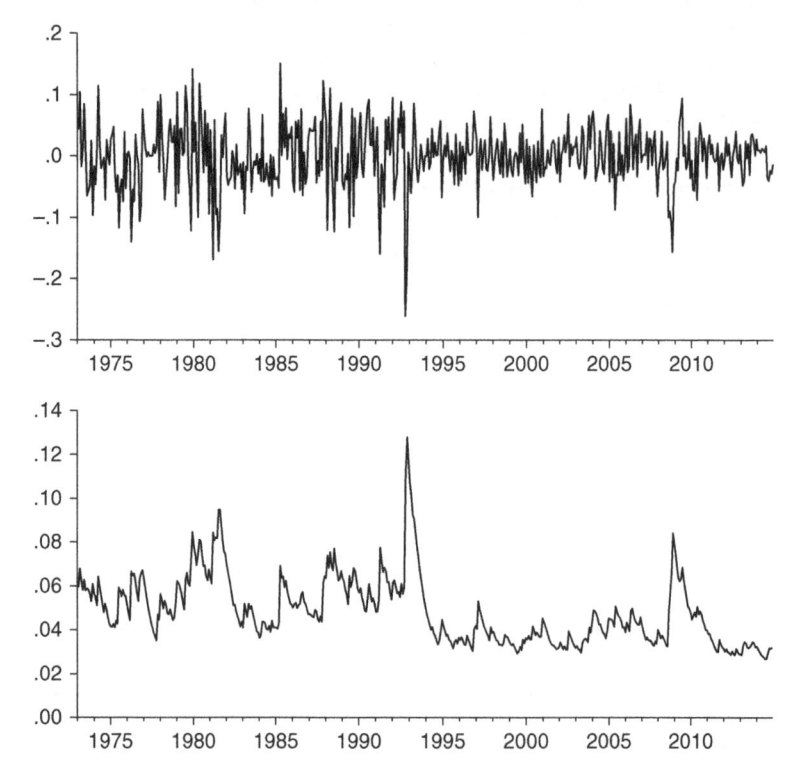

Figure 5.1 First differences of the \$/£ exchange rate (top panel). Conditional standard deviations from GARCH(1,1) model (bottom panel)

in just a few months (recall Figure 3.6). Note also the 'asymmetric' nature of $\hat{\sigma}_t$: rapid increases are followed by much slower declines, thus reflecting the persistence implied by the fitted models.

EViews Exercises

5.19 The AR(1)-GARCH(0,0) regression reported in Table 5.1 may be obtained by opening the `Ex _ 3 _ 2` page and issuing the standard least squares command

```
ls d(dollar) d(dollar(-1))
```

The ARCH(1) test statistic can be obtained by clicking *View/Residual Diagnostics/Heteroskedasticity Tests...* and selecting ARCH as the test type.

The AR(1)-GARCH(1,1) regression can then be estimated by clicking *Estimate* and selecting *ARCH – Autoregressive Conditional Heteroskedasticity* as the estimation method. The conditional standard deviation plot shown in Figure 5.1 is then obtained by clicking *View/Garch Graph/Conditional Standard Deviation* and freezing the graph.

Notes

1. The ARCH model was introduced by Robert F. Engle, 'Autoregressive conditional heteroskedasticity with estimates of the variance of UK inflation', *Econometrica* 50 (1982), 987–1008, and Engle, 'Estimates of the variance of UK inflation based on the ARCH model', *Journal of Money, Credit and Banking* 15 (1983), 286–301.
2. The GARCH model was introduced by Tim Bollerslev, 'Generalised autoregressive conditional heteroskedasticity', *Journal of Econometrics* 31 (1986), 307–27.
3. This approach was proposed by Engle, 'Autoregressive conditional heteroskedasticity'.
4. See Bollerslev, 'On the correlation structure for the generalised autoregressive conditional heteroskedastic process', *Journal of Time Series Analysis* 8 (1988), 121–32.
5. See, for example, G. William Schwert, 'Why does stock market volatility change over time?', *Journal of Finance* 44 (1989), 1115–53.
6. See Zhuanxing Ding, Granger and Engle, 'A long memory property of stock returns and a new model', *Journal of Empirical Finance* 1 (1993), 83–106.
7. See Daniel B. Nelson, "Conditional heteroskedasticity in stock returns', *Econometrica* 59 (1991), 347–70.

8. See, respectively, Matthew L. Higgins and Anil K. Bera, 'A class of nonlinear ARCH models', *International Economic Review* 33 (1992), 137–58; Jean-Michael Zakoian, 'Threshold heteroskedastic models', *Journal of Economic Dynamics and Control* 18 (1994), 931–55; and Lawrence R. Glosten, Ravi Jegannathan and David E. Runkle, 'Relationship between the expected value and the volatility of the nominal excess return on stocks', *Journal of Finance* 48 (1993), 1779–1801.
9. See Engle and G.G.J. Lee, 'A permanent and transitory component model of stock return volatility', in Engle and Halbert White (editors), *Cointegration, Causality and Forecasting: a Festschrift in Honor of Clive W.J. Granger* (Oxford University Press, 1999), 475–97.

6
Forecasting with Univariate Models

Forecasting with ARIMA models

6.1 A very important use of time series models is in forecasting. To be more precise, given a realisation $(x_{1-d}, x_{2-d}, \ldots, x_T)$ from a general ARIMA(p,d,q) process

$$\phi(B)\Delta^d x_t = \theta_0 + \theta(B)a_t \qquad (6.1)$$

how do we forecast a future value x_{T+h}?[1,2] If we let

$$\alpha(B) = \phi(B)\Delta^d = \left(1 - \alpha_1 B - \alpha_2 B^2 - \ldots - \alpha_{p+d} B^{p+d}\right)$$

(6.1) becomes, for time $T+h$,

$$\alpha(B) x_{T+h} = \theta_0 + \theta(B) a_{T+h}$$

or, when written out fully,

$$x_{T+h} = \alpha_1 x_{T+h-1} + \alpha_2 x_{T+h-2} + \ldots + \alpha_{p+d} x_{T+h-p-d} + \theta_0 + a_{T+h} \\ - \theta_1 a_{T+h-1} - \ldots - \theta_q a_{T+h-q}$$

Clearly, observations from $T+1$ onwards will be unavailable, but a *minimum mean square error* (MMSE) forecast of x_{T+h} made at time T, which we shall denote $f_{T,h}$, is given by the conditional expectation

$$f_{T,h} = E(\alpha_1 x_{T+h-1} + \alpha_2 x_{T+h-2} + \ldots + \alpha_{p+d} x_{T+h-p-d} + \theta_0$$
$$+ a_{T+h} - \theta_1 a_{T+h-1} - \ldots - \theta_q a_{T+h-q} \mid x_T, x_{T-1}, \ldots) \tag{6.2}$$

Now

$$E(x_{T+j} \mid x_T, x_{T-1}, \ldots) = \begin{cases} x_{T+j}, & j \leq 0 \\ f_{T,j}, & j > 0 \end{cases}$$

and

$$E(a_{T+j} \mid x_T, x_{T-1}, \ldots) = \begin{cases} a_{T+j}, & j \leq 0 \\ 0, & j > 0 \end{cases}$$

so that, to evaluate $f_{T,h}$, all we need to do is: (i) replace past expectations ($j \leq 0$) by known values, x_{T+j} and a_{T+j}, and (ii) replace future expectations ($j > 0$) by forecast values, $f_{T,j}$ and 0.

6.2 Three examples will illustrate the procedure. Consider first the AR(2) model $(1 - \phi_1 B - \phi_2 B^2) x_t = \theta_0 + a_t$, so that $\alpha(B) = (1 - \phi_1 B - \phi_2 B^2)$. Here

$$x_{T+h} = \phi_1 x_{T+h-1} + \phi_2 x_{T+h-2} + \theta_0 + a_{T+h}$$

and hence, for $h = 1$, we have

$$f_{T,1} = \phi_1 x_T + \phi_2 x_{T-1} + \theta_0$$

for $h = 2$

$$f_{T,2} = \phi_1 f_{T,1} + \phi_2 x_T + \theta_0$$

and, for $h > 2$,

$$f_{T,h} = \phi_1 f_{T,h-1} + \phi_2 f_{T,h-2} + \theta_0$$

An alternative expression for $f_{T,h}$ can be obtained by noting that

$$f_{T,h} = (\phi_1 + \phi_2)f_{T,h-1} - \phi_2(f_{T,h-1} - f_{T,h-2}) + \theta_0$$

from which, by repeated substitution, we may obtain

$$f_{T,h} = (\phi_1 + \phi_2)^h x_T - \phi_2 \sum_{j=0}^{h-1} (\phi_1 + \phi_2)^j (f_{T,h-1-j} - f_{T,h-2-j}) + \theta_0 \sum_{j=0}^{h-1} (\phi_1 + \phi_2)^j$$

where, by convention, we take $f_{T,0} = x_T$ and $f_{T,-1} = x_{T-1}$. Thus, for stationary processes ($\phi_1 + \phi_2 < 1$, $|\phi_2| < 1$), as the forecast horizon, or lead time, $h \to \infty$,

$$f_{T,h} \to \frac{\theta_0}{1-\phi_1-\phi_2} = E(x_t) = \mu$$

so that for long lead times the best forecast of a future observation is eventually the mean of the process.

6.3 Next consider the ARIMA(0,1,1) model $\Delta x_t = (1 - \theta B)a_t$. Here $\alpha(B) = (1 - B)$ and so

$$x_{T+h} = x_{T+h-1} + a_{T+h} - \theta a_{T+h-1}$$

For $h = 1$, we have

$$f_{T,1} = x_T - \theta a_T$$

for $h = 2$

$$f_{T,2} = f_{T,1} = x_T - \theta a_T$$

and, in general,

$$f_{T,h} = f_{T,h-1} \quad h > 1$$

Thus, for all lead times, the forecasts from origin T will follow a straight line parallel to the time axis and passing through $f_{T,1}$. Note that, since

$$f_{T,h} = x_T - \theta a_T$$

and

$$a_T = (1-B)(1-\theta B)^{-1} x_T$$

the h-step ahead forecast can be written as

$$f_{T,h} = (1-\theta)(1-\theta B)^{-1} x_T$$
$$= (1-\theta)(x_T + \theta x_{T-1} + \theta^2 x_{T-2} + \ldots)$$

so that the forecast for all future values of x is an exponentially weighted moving average of current and past values.

6.4 Finally, consider the ARIMA(0,2,2) model $\Delta^2 x_t = (1 - \theta_1 B - \theta_2 B^2) a_t$, with $\alpha(B) = (1-B)^2 = (1 - 2B + B^2)$:

$$x_{T+h} = 2x_{T+h-1} - x_{T+h-2} + a_{T+h} - \theta_1 a_{T+h-1} - \theta_2 a_{T+h-2}$$

For $h = 1$, we have

$$f_{T,1} = 2x_T - x_{T-1} - \theta_1 a_T - \theta_2 a_{T-1}$$

for $h = 2$,

$$f_{T,2} = 2f_{T,1} - x_T - \theta_1 a_T$$

for $h = 3$,

$$f_{T,3} = 2f_{T,2} - f_{T,1}$$

and thus, for $h \geq 3$,

$$f_{T,h} = 2f_{T,h-1} - f_{T,h-2}$$

Hence, for all lead times, the forecasts from origin T will follow a straight line passing through the forecasts $f_{T,1}$ and $f_{T,2}$.

Forecast errors

6.5 The h-step ahead forecast error for origin T is

$$e_{T,h} = x_{T+h} - f_{T,h} = a_{T+h} + \psi_1 a_{T+h-1} + \ldots + \psi_{h-1} a_{T+1}$$

where $\psi_1,\ldots,\psi_{h-1}$ are the first $h-1$ ψ-weights in $\psi(B) = \alpha^{-1}(B)\theta(B)$. The variance of this forecast error is then

$$V(e_{T,h}) = \sigma^2(1 + \psi_1^2 + \psi_2^2 + \ldots + \psi_{h-1}^2) \tag{6.3}$$

The forecast error is therefore a linear combination of the unobservable future shocks entering the system after time T and, in particular, the one-step ahead forecast error will be

$$e_{T,1} = x_{T+1} - f_{T,1} = a_{T+1}$$

Thus, for a MMSE forecast, the one-step ahead forecast errors must be uncorrelated. However, h-step ahead forecasts made at different origins will not be uncorrelated, and neither will be forecasts for different lead times made at the same origin.[3]

6.6 For the AR(2) model, we have $\psi_1 = \phi_1, \psi_2 = \phi_1^2 + \phi_2$ and, for $j > 2$, $\psi_j = \phi_1 \psi_{j-1} + \phi_2 \psi_{j-2}$ (recall §2.13). Since we are assuming stationarity, these ψ-weights converge absolutely, which implies that $\sum_{j=1}^{h} \psi_j^2 < \infty$. Consequently $V(e_{T,h})$ converges to a finite value, which is the variance of the process about the ultimate forecast μ.

For the ARIMA(0,1,1) model, $\psi_j = 1 - \theta, j = 1,2,\ldots$ Thus we have

$$V(e_{T,h}) = \sigma^2(1 + (h-1)(1-\theta)^2)$$

which increases with h. Similarly, the ARIMA(0,2,2) model has ψ-weights given by $\psi_j = 1 + \theta_2 + j(1 - \theta_1 - \theta_2)$, $j = 1,2,\ldots$, and an h-step ahead forecast error variance of

$$V(e_{T,h}) = \sigma^2(1 + (h-1)(1+\theta_2)^2 + \tfrac{1}{6}h(h-1)(2h-1)(1-\theta_1-\theta_2)^2 + h(h-1)(1+\theta_2)(1-\theta_1-\theta_2))$$

which again increases with h but potentially more rapidly than the ARIMA(0,1,1).

These examples show how the degree of differencing (equivalently, the order of integration) determines not only how successive forecasts are related to each other, but also the behaviour of the associated error variances.

EXAMPLE 6.1 ARIMA forecasting of the spread

Example 2.2 fitted an AR(2) model to the UK interest rate spread, yielding parameter estimates $\phi_1 = 1.192$, $\phi_2 = -0.224$ and $\theta_0 = 0.035$. With the last two observations being $x_{T-1} = 2.31$ (November 2014) and $x_T = 2.06$ (December 2014), forecasts are obtained as

$f_{T,1} = 1.192 x_T - 0.224 x_{T-1} + 0.035 = 1.975$ (January 2015)

$f_{T,2} = 1.192 f_{T,1} - 0.224 x_T + 0.035 = 1.929$ (February 2015)

$f_{T,3} = 1.192 f_{T,2} - 0.224 f_{T,1} + 0.035 = 1.894$ (March 2015)

and so on. As h increases, the forecasts eventually tend to 1.174, the sample mean of the spread, although the large autoregressive root makes this convergence to the sample mean rather slow. The ψ-weights are given by

$\psi_1 = \phi_1 = 1.192$

$\psi_2 = \phi_1^2 + \phi_2 = 1.197$

$\psi_3 = \phi_1^3 + 2\phi_1\phi_2 = 1.160$

$$\psi_4 = \phi_1^4 + 3\phi_1^2\phi_2 + \phi_2^2 = 1.114$$

and, hence,

$$\psi_h = 1.192\psi_{h-1} - 0.224\psi_{h-2}$$

With $\sigma = 0.401$, the forecast error variances are thus

$$V(e_{T,1}) = 0.401^2 = 0.161$$

$$V(e_{T,2}) = 0.401^2 \left(1 + 1.192^2\right) = 0.389$$

$$V(e_{T,3}) = 0.401^2 \left(1 + 1.192^2 + 1.197^2\right) = 0.620$$

$$V(e_{T,4}) = 0.401^2 \left(1 + 1.192^2 + 1.197^2 + 1.160^2\right) = 0.836$$

$$\vdots$$

these eventually converging to the sample variance of the spread, 3.406.

If, however, we use the ARIMA(0,1,1) process of Example 3.1 to model the spread, with $\hat{\theta} = 0.204$ and $\hat{\sigma} = 0.405$ (conveniently setting the insignificant drift to 0), then our forecasts are (using the December 2014 residual $\hat{a}_T = -0.236$)

$$f_{T,1} = 2.06 + 0.204 \times 0.236 = 2.012$$

and, for $h > 1$,

$$f_{T,h} = f_{T,1} = 2.012$$

so that there is no tendency for the forecasts to converge to the sample mean or, indeed, to any other value. Furthermore, the forecast error variances are given by

$$V(e_{T,h}) = 0.405^2(1 + 0.634(h-1)) = 0.164 + 0.104(h-1)$$

which, of course, increase with h, rather than tending to a constant. This example thus illustrates, within a forecasting context, the radically different properties of ARMA models which have, on the one hand, a unit autoregressive root and, on the other, a root that is large but less than unity.

Forecasting a trend stationary process

6.7 Let us now consider the TS process

$$x_t = \beta_0 + \beta_1 t + \varepsilon_t \qquad \phi(B)\varepsilon_t = \theta(B)a_t \tag{6.4}$$

The forecast of x_{T+h} made at time T is

$$f_{T,h} = \beta_0 + \beta_1(T+h) + g_{T,h}$$

where $g_{T,h}$ is the forecast of ε_{T+h}, which from (6.2) is given by

$$g_{T,h} = E\begin{pmatrix} \phi_1\varepsilon_{T+h-1} + \phi_2\varepsilon_{T+h-2} + \ldots + \phi_p\varepsilon_{T+h-p} + a_{T+h} - \theta_1 a_{T+h-1} \\ -\ldots - \theta_q a_{T+h-q} \mid \varepsilon_T, \varepsilon_{T-1}, \ldots \end{pmatrix}$$

Since ε_t is, by assumption, stationary, $g_{T,h} \to 0$ as $h \to \infty$. Thus, for large h, $f_{T,h} \to \beta_0 + \beta_1(T+h)$ and forecasts will be given simply by the extrapolated linear trend. For smaller h there will also be the component $g_{T,h}$, but this will decrease in size as h increases. The forecast error will be

$$e_{T,h} = x_{T+h} - f_{T,h} = \varepsilon_{T+h} - g_{T,h}$$

and hence the uncertainty in any TS forecast is due solely to the error in forecasting the ARMA component. As a consequence, the forecast error variance is bounded by the sample variance of ε_t, and this is in sharp contrast to the forecast error variance of the ARIMA(p,2,q) process, which, from §6.4, also has forecasts that lie on a (different) linear trend, but these have *unbounded* error variances. In the simplest case in which ε_t is white noise, *all* forecasts of a TS process have the *same* error variance, σ^2.

EXAMPLE 6.2 Forecasting the *FTA All Share* index as a TS process

A TS model fitted to (the logarithms of) the *FTA All Share* index was estimated to be

$$x_t = 3.905 + 0.0060t + \varepsilon_t$$

$$\varepsilon_t = 1.111\varepsilon_{t-1} - 0.123\varepsilon_{t-2} + a_t$$

That this is a misspecified model is clear from Figure 6.1, which superimposes the fitted linear trend and reveals that there are highly persistent deviations of the series from the trend, confirmed by the largest autoregressive root being estimated to be 0.99. The artificiality of this example notwithstanding, very long horizon forecasts of x_{T+h} would then be given by $f_{T,h} = 3.905 + 0.0060(T + h)$, although shorter horizon forecasts will have appended the forecast of the AR(2) component, $g_{T,h} = 1.111 g_{T,h-1} - 0.123\ g_{T,h-2}$. Because of the (very near) non-stationarity of this component, $g_{T,h}$ will decline only very slowly to 0, so that $f_{T,h}$ will continue to lie well below the extrapolated trend for h reasonably large: the December 2015 ($h = 12$) forecast is $f_{T,12} = 8.27$ compared to the extrapolated trend of 8.53.

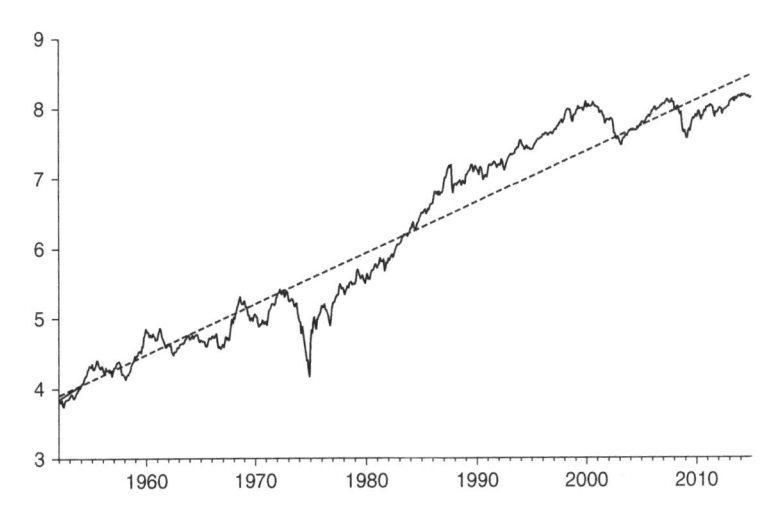

Figure 6.1 Logarithms of the *FTA All Share* index with linear trend superimposed

Forecasting from an ARMA-GARCH model

6.8 Suppose we have the ARMA(P,Q)-GARCH(p,q) model of §5.8,

$$x_t = \Phi_1 x_{t-1} + \ldots + \Phi_P x_{t-P} + \Theta_0 + \varepsilon_t - \Theta_1 \varepsilon_{t-1} - \ldots - \Theta_Q \varepsilon_{t-Q} \quad (6.5)$$

$$\sigma_t^2 = \alpha_0 + \alpha_1 \varepsilon_{t-1}^2 + \ldots + \alpha_p \varepsilon_{t-p}^2 + \beta_1 \sigma_{t-1}^2 + \ldots + \beta_q \sigma_{t-q}^2 \quad (6.6)$$

Forecasts of x_{T+h} can be obtained from the 'mean equation' (6.5) in the manner outlined in §§6.1-6.4. When calculating forecast error variances, however, it can no longer be assumed that the error variance itself is constant. Thus (6.2) must be amended to

$$V(e_{t,h}) = \sigma_{T+h}^2 + \psi_1^2 \sigma_{T+h-1}^2 + \ldots + \psi_{h-1}^2 \sigma_{T+1}^2$$

with the σ_{T+h}^2 obtained recursively from (6.6).

EXAMPLE 6.3 Forecasting the $/£ exchange rate

In Example 5.1 we found that this exchange rate could be modelled as

$$x_t = x_{t-1} + \varepsilon_t$$
$$\sigma_t^2 = 0.000083 + 0.144 \varepsilon_{t-1}^2 + 0.828 \sigma_{t-1}^2$$

Forecasts of the exchange rate are thus given by $f_{T,h} = 1.564$ for all h, this being the December 2014 rate. Since a pure random walk has $\psi_i = 1$ for all i, the forecast error variances are given by

$$V(e_{t,h}) = \sigma_{T,h}^2 + \sigma_{T,h-1}^2 + \ldots + \sigma_{T,1}^2$$

where, using the December 2014 residual $e_T = -0.014$ and conditional error variance $\hat{\sigma}_T^2 = 0.001043$,

$$\sigma_{T,1}^2 = 0.000083 + 0.144 e_T^2 + 0.828 \hat{\sigma}_T^2$$
$$= 0.000083 - 0.144 \times 0.014^2 + 0.828 \times 0.001043$$
$$= 0.000975$$

$$\sigma_{T,j}^2 = 0.000083 + 0.828 \sigma_{T,j-1}^2 \quad j \geq 2$$

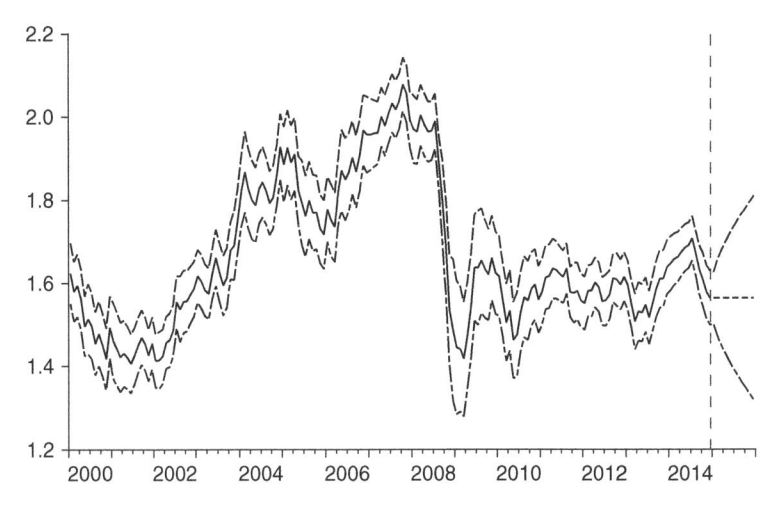

Figure 6.2 $/£ exchange rate from 2000 onwards with 2 conditional standard error bounds and forecasts out to December 2015

Figure 6.2 shows the exchange rate enclosed by 2 conditional standard error bounds for 2000 onwards and with forecasts from January 2015 to December 2015. Note how the width of the bounds interval varies through time, most notably increasing during the large fall in the exchange rate during the financial crisis from August 2008 to April 2009. Also note how the conditional standard error bounds increase rapidly in the forecast period, so that the forecasts quickly become very imprecise: the December 2015 forecast 'interval' (which is approximately a 95% one) is 1.32–1.80 $/£.

EViews Exercises

6.9 The competing forecasts of the spread may be computed using the variable spread given in page Ex_6_1, which extends the sample range out to December 2015 to accommodate the forecasts for 2015. The AR(2) forecasts are obtained from the regression

```
ls spread c spread(-1 to -2)
```

by clicking **Forecast**, selecting the forecast sample as 2015m01 2015m12 and, optionally, changing the forecast name to, say, spread _ ar and unchecking 'Insert actuals for out-of-sample observations'. Inserting ar _ se, say, into the 'S.E. (optional) box' will place standard errors for the forecast sample into this variable.

The forecasts from the ARIMA(0,1,1) model, estimated by ls d(spread) ma(1), can be obtained by repeating the procedure but ensuring that spread is selected as the series to be forecast, optionally changing the forecast and standard errors names to, say, spread _ ma and ma _ se.

6.10 The forecasts for the TS model of the *FTA All Share* index in Example 6.2 may be calculated using the logarithms of the index, the variable p of Ex _ 6 _ 2. The TS model can be estimated with the command

```
ls p c @trend ar(1) ar(2)
```

The extrapolated trend, p _ trend say, can be obtained by forecasting over the entire sample 1952m01 2015m12 on selecting the 'Dynamic Forecast' method. The forecasts $f_{T,h}$, p _ f say, can be obtained by selecting the 'Static Forecast' method, but this only gives the forecast $f_{T,1}$, the January 2015 (2015m01) forecast. The remaining forecasts can be computed with the commands

```
smpl 2015m02 2015m12
genr p _ f = p _ trend + 1.1112*(p _ f(-1) - p _ trend(-1))
            -0.1229*(p _ f(-2) - p _ trend(-2))
```

6.11 In constructing Figure 6.2, the conditional standard error bounds for 2000 to 2014 can be obtained by estimating the GARCH(1,1) model for d(dollar) as in §5.19 and clicking **Proc/Make GARCH Variance Series**... and saving the series as, say, garch01. Two conditional standard error bounds may then be computed as

```
genr dollar _ upp = dollar + 2*sqr(garch01)
genr dollar _ low = dollar - 2*sqr(garch01)
```

The forecasts and forecast standard errors for 2015 may be obtained by computing a dynamic forecast for dollar for the period 2015m01 2015m12. Upper and lower bounds may then be constructed as above and the various series put together to form the graph shown as Figure 6.2.

Notes

1. A detailed exposition of forecasting from ARIMA models is provided by Box and Jenkins, *Time Series Analysis*, Chapter 5. A wide ranging discussion of forecasting economic time series is to be found in Granger and Newbold, *Forecasting Economic Time Series*.
2. Throughout this chapter we use 'forecast' rather than 'predict', even though they are commonly regarded as synonyms. This is because the modern literature on the econometrics of forecasting defines the two terms differently, although the difference is subtle and rather deep. For a detailed discussion of these definitions, see Michael P. Clements and David F. Hendry, *Forecasting Economic Time Series* (Cambridge University Press, 1998; chapter 2). Briefly, predictability is defined to be a property of a random variable in relation to an information set (the conditional and unconditional distributions of the variable do not coincide). It is a necessary, but not sufficient, condition for forecastability, as the latter also requires knowledge of what information is relevant for forecasting and how it enters the causal mechanism.
3. See, for example, Box and Jenkins, *Time Series Analysis*, appendix A5.1.

7
Modelling Multivariate Time Series: Vector Autoregressions and Granger Causality

Dynamic regression models

7.1 So far our focus has just been on modelling individual time series but we now extend the analysis to multivariate models. To develop methods of modelling a vector of time series, consider again the AR(1) process, now written for the stationary series y_t and with a slightly different notation to that used before:

$$y_t = \theta + \phi y_{t-1} + a_t \tag{7.1}$$

The standard *dynamic regression model* adds exogenous variables, perhaps with lags, to the right-hand side of (7.1); to take the simplest example of a single exogenous variable x_t having a single lag, consider

$$y_t = c + a y_{t-1} + b_0 x_t + b_1 x_{t-1} + e_t \tag{7.2}$$

Again, note the change of notation as coefficients and innovations will not, in general, be the same across (7.1) and (7.2): c and a will differ from θ and ϕ, as will the variance of e_t, σ_e^2, differ from that of a_t, σ_a^2, with, typically, $\sigma_e^2 < \sigma_a^2$ if the additional coefficients b_0 and b_1 are non-zero.

7.2 Now suppose that we have *two* endogenous variables, $y_{1,t}$ and $y_{2,t}$, that may both be related to x_t and its lags and also to lags of each other; again, in the simplest case

$$y_{1,t} = c_1 + a_{11}y_{1,t-1} + a_{12}y_{2,t-1} + b_{10}x_t + b_{11}x_{t-1} + u_{1,t} \qquad (7.3)$$

$$y_{2,t} = c_2 + a_{21}y_{1,t-1} + a_{22}y_{2,t-1} + b_{20}x_t + b_{21}x_{t-1} + u_{2,t}$$

The 'system' contained in equation (7.3) is known as the **multivariate dynamic regression model**.[1] Note that the 'contemporaneous' variables, $y_{1,t}$ and $y_{2,t}$, are not included as regressors in the equations for $y_{2,t}$ and $y_{1,t}$, respectively, as this would lead to simultaneity and an identification problem, in the sense that the two equations making up (7.3) would then be statistically indistinguishable, there being the same variables in both. Of course, $y_{1,t}$ and $y_{2,t}$ may well be contemporaneously correlated, and any such correlation can be modelled by allowing the covariance between the innovations to be non-zero, so that $E(u_{1,t}u_{2,t}) = \sigma_{12}$ say, the variances of the two innovations being $E(u_1^2) = \sigma_1^2$ and $E(u_2^2) = \sigma_2^2$.

Vector autoregressions

7.3 The pair of equations in (7.3) may be generalised to a model containing n endogenous variables and k exogenous variables.[2] Gathering these together in the vectors $\mathbf{y}_t' = (y_{1,t}, y_{2,t}, \ldots, y_{n,t})$ and $\mathbf{x}_t' = (x_{1,t}, x_{2,t}, \ldots, x_{k,t})$, the general form of the model may be written as

$$\mathbf{y}_t = \mathbf{c} + \sum_{i=1}^{p} \mathbf{A}_i \mathbf{y}_{t-i} + \sum_{i=0}^{q} \mathbf{B}_i \mathbf{x}_{t-i} + \mathbf{u}_t \qquad (7.4)$$

where there are a maximum of p lags on the endogenous variables and a maximum of q lags on the exogenous variables. Here $\mathbf{c}' = (c_1, c_2, \ldots, c_n)$ is a $1 \times n$ vector of constants and $\mathbf{A}_1, \mathbf{A}_2, \ldots, \mathbf{A}_p$ and $\mathbf{B}_0, \mathbf{B}_1, \mathbf{B}_2, \ldots, \mathbf{B}_q$ are sets of $n \times n$ and $n \times k$ matrices of regression coefficients, respectively, such that

$$\mathbf{A}_i = \begin{bmatrix} a_{11,i} & a_{12,i} & \cdots & a_{1n,i} \\ a_{21,i} & a_{22,i} & \cdots & a_{2n,i} \\ \vdots & & & \vdots \\ a_{n1,i} & a_{n2,i} & \cdots & a_{nn,i} \end{bmatrix} \qquad \mathbf{B}_i = \begin{bmatrix} b_{11,i} & b_{12,i} & \cdots & b_{1k,i} \\ b_{21,i} & b_{22,i} & \cdots & b_{2k,i} \\ \vdots & & & \vdots \\ b_{n1,i} & b_{n2,i} & \cdots & b_{nk,i} \end{bmatrix}$$

$\mathbf{u}'_t = (u_{1,t}, u_{2,t}, \ldots, u_{n,t})$ is a $1 \times n$ vector of innovations (or errors), whose variances and covariances can be gathered together in the symmetric $n \times n$ *error covariance* matrix

$$\Omega = E(\mathbf{u}_t \mathbf{u}'_t) = \begin{bmatrix} \sigma_1^2 & \sigma_{12} & \cdots & \sigma_{1n} \\ \sigma_{12} & \sigma_2^2 & \cdots & \sigma_{2n} \\ \vdots & & & \vdots \\ \sigma_{1n} & \sigma_{2n} & \cdots & \sigma_n^2 \end{bmatrix}$$

It is assumed that these errors are mutually uncorrelated, so that $E(\mathbf{u}_t \mathbf{u}'_s) = \mathbf{0}$ for $t \neq s$, where $\mathbf{0}$ is an $n \times n$ null matrix.

7.4 The model (7.4) may be estimated by (multivariate) least squares if there are *exactly* p lags of the endogenous variables and q lags of the exogenous variables in each equation. If there are different lag lengths in individual equations then a systems estimator needs to be used to obtain efficient estimates.[3]

7.5 Suppose the model (7.4) does not contain any exogenous variables, so that all the $\mathbf{B}_i$ matrices are $\mathbf{0}$, and that there are p lags of the endogenous variables in *every* equation:

$$\mathbf{y}_t = \mathbf{c} + \sum_{i=1}^{p} \mathbf{A}_i \mathbf{y}_{t-i} + \mathbf{u}_t \tag{7.5}$$

Because (7.5) is now simply a pth order autoregression in the vector $\mathbf{y}_t$ it is known as a *vector autoregression* (VAR(p)) of dimension n and again can be estimated by multivariate least squares.[4] VARs have become extremely popular for modelling multivariate systems of economic and financial time series because the absence of $\mathbf{x}_t$ terms precludes having to make any endogenous-exogenous classification of the variables, for such distinctions are often considered to be highly contentious.

Granger causality

7.6 In the VAR (7.5) the presence of non-zero off-diagonal elements in the $\mathbf{A}_i$ matrices, $a_{rs,i} \neq 0$, $r \neq s$, implies that there are dynamic relationships between the variables, otherwise the model would collapse

Modelling Multivariate Time Series 101

to a set of n univariate AR processes. The presence of such dynamic relationships is known as **Granger (–Sims) causality**.[5] The variable y_s does not Granger-cause the variable y_r if $a_{rs,i} = 0$ for all $i = 1,2,...,p$. If, on the other hand, there is at least one $a_{rs,i} \neq 0$ then y_s is said to Granger-cause y_r because if that is the case then past values of y_s are useful in forecasting the current value of y_r: Granger-causality is thus a criterion of 'forecastability'. If y_r also Granger-causes y_s, the pair of variables are said to exhibit **feedback**.

7.7 The presence of non-zero off-diagonal elements in the error covariance matrix Ω signals the presence of simultaneity. For example, $\sigma_{rs} \neq 0$ implies that $y_{r,t}$ and $y_{s,t}$ are contemporaneously correlated. It might be tempting to try and model such correlation by including $y_{r,t}$ in the equation for $y_{s,t}$ but, if this is done, then $y_{s,t}$ could equally well be included in the $y_{r,t}$ equation. As was pointed out in §7.2, this would lead to an identification problem, since the two equations would be statistically indistinguishable and the VAR could no longer be estimated. The presence of $\sigma_{rs} \neq 0$ is sometimes referred to as **instantaneous causality**, although we should be careful when interpreting this phrase, as *no* causal direction can be inferred from σ_{rs} being non-zero (recall the 'correlation does not imply causation' argument found in any basic statistics text).[6]

Determining the lag order of a VAR and testing for causality

7.8 To enable the VAR to become operational the lag order p, which will typically be unknown, has to be determined empirically. A traditional way of selecting the lag order is to use a sequential testing procedure. Consider the model (7.5) with error covariance matrix $\Omega_p = E(\mathbf{u}_t \mathbf{u}_t')$, where a p subscript is included to emphasise that the matrix is related to a VAR(p). An estimate of this matrix is given by

$$\hat{\Omega}_p = (T - p)^{-1} \hat{\mathbf{U}}_p \hat{\mathbf{U}}_p'$$

where $\hat{\mathbf{U}}_p = (\hat{\mathbf{u}}_1',...,\hat{\mathbf{u}}_n')'$ is the matrix of residuals obtained by OLS estimation of the VAR(p), $\hat{\mathbf{u}}_r = (\hat{u}_{r,p+1},...,\hat{u}_{r,T})'$ being the residual vector from the rth equation (noting that with a sample of size T,

p observations will be lost through lagging). A likelihood ratio (LR) statistic for testing the order p against the order m, $m < p$, is

$$LR(p,m) = (T - np)\log\left(\left|\hat{\Omega}_m\right|/\left|\hat{\Omega}_p\right|\right) \sim \chi^2_{n^2(p-m)} \qquad (7.6)$$

Thus if $LR(p,m)$ exceeds the α critical value of the χ^2 distribution with $n^2(p - m)$ degrees of freedom then the hypothesis that the VAR order is m is rejected at the α level in favour of the higher order p. The statistic uses the scaling factor $T - np$ rather than $T - p$ to account for possible small-sample bias.

The statistic (7.6) may then be used sequentially beginning with a maximum value of p, p_{max} say, testing first p_{max} against $p_{max} - 1$ using $LR(p_{max}, p_{max} - 1)$ and, if this statistic is not significant, then testing $p_{max} - 1$ against $p_{max} - 2$ using $LR(p_{max} - 1, p_{max} - 2)$, continuing until a significant test is obtained.

7.9 Alternatively, some type of information criterion can be minimised. These are essentially multivariate extensions of those introduced in Example 2.3: for example, the multivariate AIC and BIC criteria are defined as

$$MAIC(p) = \log\left|\hat{\Omega}_p\right| + (2 + n^2 p)T^{-1}$$

$$MBIC(p) = \log\left|\hat{\Omega}_p\right| + n^2 pT^{-1}\ln T \qquad p = 0, 1, \ldots, p_{max}$$

7.10 After an order has been selected and the VAR fitted, checks on its adequacy need to be performed. There are analogues to the diagnostic checks used for univariate models and introduced in the examples of Chapter 2, but with vector time series there is probably no substitute for detailed inspection of the residual correlation structure, including cross-correlations, for revealing subtle relationships that may indicate important directions for model improvement.

EXAMPLE 7.1 The interaction of the UK bond and gilt markets

The VAR framework requires that the time series making up the vector $\mathbf{y}_t$ be stationary. Example 4.3 demonstrated that short and long UK interest rates are $I(1)$, so that their differences, ΔRS_t and $\Delta R20_t$, will be stationary. Since these series may be thought of as being representative of the bond and gilt markets, respectively, the interaction between the two markets may be investigated by first determining the order of the two-dimensional VAR for $\mathbf{y}_t = (\Delta RS_t, \Delta R20_t)'$. Table 7.1 shows various statistics for doing this for a maximum setting of $p_{\max} = 4$. The LR and $MAIC$ statistics select an order of 2 while the $MBIC$ selects an order of 1, although the VAR(1) fit leaves a significant second order residual autocorrelation in the $\Delta R20$ equation. An order of 2 was therefore chosen, with the fitted VAR(2) being

$$\begin{bmatrix} \Delta RS_t \\ \Delta R20_t \end{bmatrix} = \begin{bmatrix} 0.217 & 0.281 \\ (0.041) & (0.063) \\ -0.011 & 0.310 \\ (0.026) & (0.041) \end{bmatrix} \begin{bmatrix} \Delta RS_{t-1} \\ \Delta R20_{t-1} \end{bmatrix} + \begin{bmatrix} 0.021 & -0.066 \\ (0.040) & (0.063) \\ 0.022 & -0.139 \\ (0.026) & (0.041) \end{bmatrix} \begin{bmatrix} \Delta RS_{t-2} \\ \Delta R20_{t-2} \end{bmatrix} + \begin{bmatrix} \hat{u}_{1,t} \\ \hat{u}_{2,t} \end{bmatrix}$$

The intercept vector $\mathbf{c}$ has been excluded from the model as, consistent with Example 4.3, it was found to be insignificant. Various checks on the residuals of the VAR(2) failed to uncover any model inadequacy.

7.11 Within a VAR(p), Granger-causality running from y_s to y_r, which may be depicted as $y_s \to y_r$, can be evaluated by setting up the null hypothesis of **non-Granger-causality** ($y_s \not\to y_r$), $H_0 : a_{rs,1} = \ldots = a_{rs,p} = 0$, and testing this with a Wald (F)-statistic, a multivariate extension of the standard F-statistic for testing a set of zero restrictions in a conventional regression model.[7]

Table 7.1 Order determination statistics for $\mathbf{y}_t = (\Delta RS_t, \Delta R20_t)'$

p	log L	LR($p,p-1$)	MAIC	MBIC
0	−539.15	−	1.441	1.453
1	−477.55	122.70	1.288	1.325*
2	−471.39	12.24*	1.282*	1.344
3	−470.63	1.51	1.291	1.377
4	−469.92	1.40	1.299	1.410

$LR(p, p-1) \sim \chi_4^2 \quad \chi_4^2(0.05) = 9.49$

EXAMPLE 7.2   Testing for Granger-causality between the UK gilt and bond markets

Within the estimated AR(2) model of Example 7.1, the Wald statistics for Granger-causality test $a_{12,1} = a_{12,2} = 0$ for the null $\Delta R20 \nrightarrow \Delta RS$, and $a_{21,1} = a_{21,2} = 0$ for the null $\Delta RS \nrightarrow \Delta R20$. These statistics are 20.13 and 0.76, respectively, and reveal that the long-rate Granger causes the short rate but that there is *no* feedback: movements in the gilt market thus lead, and so help to forecast, movements in the bond market.

Variance decompositions and innovation accounting

7.12 While the estimated coefficients of a VAR(1) are relatively easy to interpret, this quickly becomes problematic for higher order VARs because not only does the number of coefficients increase rapidly, but many of the coefficients will be imprecisely estimated and highly inter-correlated, so becoming statistically insignificant, as can be seen in the estimated VAR(2) of Example 7.1, where only $\hat{a}_{22,2}$ in $\hat{\mathbf{A}}_2$ is significant. This has led to the development of several techniques for examining the 'information content' of a VAR that are based on the vector moving average representation of $\mathbf{y}_t$. Suppose that the VAR is written in lag operator form as

$$\mathbf{A}(B)\mathbf{y}_t = \mathbf{u}_t$$

where

$$\mathbf{A}(B) = \mathbf{I}_n - \mathbf{A}_1 B - \ldots - \mathbf{A}_p B^p$$

is a matrix polynomial in B. Analogous to the univariate case (recall §§2.7–2.8), the (infinite) vector MA representation is

$$\mathbf{y}_t = \mathbf{A}^{-1}(B)\mathbf{u}_t = \Psi(B)\mathbf{u}_t = \mathbf{u}_t + \sum_{i=1}^{\infty} \Psi_i \mathbf{u}_{t-i} \tag{7.7}$$

where

$$\Psi_i = \sum_{j=1}^{i} \mathbf{A}_j \Psi_{i-j} \qquad \Psi_0 = \mathbf{I}_n \qquad \Psi_i = \mathbf{0} \quad i < 0$$

this recursion being obtained by equating coefficients of B in $\Psi(B)\mathbf{A}(B) = \mathbf{I}_n$.

Modelling Multivariate Time Series 105

7.13 The Ψ_i matrices can be interpreted as the *dynamic multipliers* of the system, since they represent the model's response to a unit shock in each of the variables. The response of y_r to a unit shock in y_s (produced by $u_{s,t}$ taking the value unity rather than its expected value of 0) is therefore given by the *impulse response function*, which is the sequence $\psi_{rs,1}, \psi_{rs,2}, \ldots$, where $\psi_{rs,i}$ is the r,sth element of the matrix Ψ_i. Since $\Omega_p = E(\mathbf{u}_t\mathbf{u}_t')$ is not required to be diagonal, the components of $\mathbf{u}_t$ are allowed to be contemporaneously correlated. If these correlations are high, simulation of a shock to y_s, while all other components of $\mathbf{u}_t$ are held constant, could be misleading, as there is no way of separating out the response of y_r to a y_s shock from its response to other shocks that are correlated with $u_{s,t}$. However, if we define the lower triangular matrix $\mathbf{S}$ such that $\mathbf{SS}' = \Omega_p$ and define $\mathbf{v}_t = \mathbf{S}^{-1}\mathbf{u}_t$, then $E(\mathbf{v}_t\mathbf{v}_t') = \mathbf{I}_n$ and the transformed errors $\mathbf{v}_t$ are orthogonal to each other (this is known as a *Cholesky decomposition*). The MA representation can then be renormalised into the *recursive* form

$$\mathbf{y}_t = \sum_{i=0}^{\infty}(\Psi_i\mathbf{S})(\mathbf{S}^{-1}\mathbf{u}_{t-i}) = \sum_{i=0}^{\infty}\Psi_i^O\mathbf{v}_{t-i}$$

where $\Psi_i^O = \Psi_i\mathbf{S}$ (so that $\Psi_0^O = \Psi_0\mathbf{S}$ is lower triangular). The impulse response function of y_r to a y_s shock is then given by the sequence $\psi_{rs,0}^O, \psi_{rs,1}^O, \psi_{rs,2}^O, \ldots$, where each impulse response can be written compactly as

$$\psi_{rs,i}^O = \mathbf{e}_r'\Psi_i\mathbf{S}\mathbf{e}_s \qquad (7.8)$$

Here $\mathbf{e}_s$ is the $n \times 1$ selection vector containing unity as the sth element and zeroes elsewhere. This sequence is known as the *orthogonalised impulse response function*.

7.14 The uncorrelatedness of the v_ts allows the error variance of the h-step ahead forecast of y_r to be decomposed into components accounted for by these innovations, a technique thus known as *innovation accounting*.[8] In particular, the proportion of the h-step ahead forecast error variance of y_r accounted for by the orthogonalised innovations to y_s is given by

$$V_{rs,h}^O = \frac{\sum_{i=0}^{h}(\psi_{rs,h}^O)^2}{\sum_{i=0}^{h}\mathbf{e}_r'\Psi_i\Omega_p\Psi_i'\mathbf{e}_r} = \frac{\sum_{i=0}^{h}(\mathbf{e}_r'\Psi_i\mathbf{S}\mathbf{e}_s)^2}{\sum_{i=0}^{h}\mathbf{e}_r'\Psi_i\Omega_p\Psi_i'\mathbf{e}_r}$$

For large h, this **orthogonalised forecast error variance decomposition** allows the isolation of those relative contributions to variability that are, intuitively, 'persistent'. The technique of orthogonalisation does, however, have an important disadvantage, for the choice of the **S** matrix is not unique, so that different choices (most notably, different orderings of the variables) will alter the $\psi_{rs,i}^O$ coefficients and hence the impulse response functions and variance decompositions. The extent of these changes will depend on the size of the contemporaneous correlations between the innovations.[9]

7.15 Apart from comparing the impulse responses and variance decompositions for alternative orderings of the variables, one solution to this problem is to use *generalised impulse responses*, defined by replacing **S** in (7.8) with $\sigma_r^{-1}\Omega_p$:[10]

$$\psi_{rs,i}^G = \sigma_r^{-1}\mathbf{e}_r'\Psi_i\Omega_p\mathbf{e}_s$$

The generalised impulse responses are invariant to the ordering of the variables, are unique, and fully take into account the historical patterns of correlations observed amongst the different shocks. The orthogonalised and generalised impulse responses coincide only when Ω_p is diagonal, and in general are only the same for $s = 1$.

EXAMPLE 7.3 Variance decomposition and innovation accounting for the bond and gilt markets

From Example 7.1, the VAR(2) fitted to $\mathbf{y}_t = (\Delta RS_t, \Delta R20_t)'$ has

$$\hat{\mathbf{A}}_1 = \begin{bmatrix} 0.217 & 0.281 \\ -0.011 & 0.310 \end{bmatrix} \quad \hat{\mathbf{A}}_2 = \begin{bmatrix} 0.021 & -0.066 \\ 0.022 & -0.139 \end{bmatrix}$$

The vector MA representation (7.7) then has coefficient matrices given by

$$\Psi_i = \mathbf{A}_1\Psi_{i-1} + \mathbf{A}_2\Psi_{i-2}$$

Modelling Multivariate Time Series 107

so that

$$\Psi_1 = A_1\Psi_0 = \begin{bmatrix} 0.217 & 0.281 \\ -0.011 & 0.310 \end{bmatrix}$$

$$\Psi_2 = A_1\Psi_1 + A_2\Psi_0 = \begin{bmatrix} 0.217 & 0.281 \\ -0.011 & 0.310 \end{bmatrix}^2 + \begin{bmatrix} 0.021 & -0.066 \\ 0.022 & -0.139 \end{bmatrix}$$

$$= \begin{bmatrix} 0.065 & 0.082 \\ 0.016 & -0.046 \end{bmatrix}$$

$$\Psi_3 = A_1\Psi_2 + A_2\Psi_1 = \begin{bmatrix} 0.024 & -0.010 \\ 0.011 & -0.052 \end{bmatrix}$$

$\vdots$

The estimated error covariance matrix is

$$\hat{\Omega}_2 = \begin{bmatrix} 0.190 & 0.056 \\ 0.056 & 0.080 \end{bmatrix}$$

so that the contemporaneous correlation between the innovations is 0.46, thus necessitating orthogonalisation. The Cholesky decomposition of $\hat{\Omega}_2$ for the ordering ΔRS, $\Delta R20$ is

$$S = \begin{bmatrix} 0.436 & 0 \\ 0.129 & 0.252 \end{bmatrix} = \Psi_0^O$$

with

$$S^{-1} = \begin{bmatrix} 2.283 & 0 \\ -1.173 & 3.955 \end{bmatrix}$$

Thus

$$\Psi_1^O = \Psi_1 S = \begin{bmatrix} 0.131 & 0.071 \\ 0.035 & 0.078 \end{bmatrix}$$

$$\Psi_2^O = \Psi_2 S = \begin{bmatrix} 0.039 & 0.021 \\ 0.001 & -0.012 \end{bmatrix}$$

$$\Psi_3^O = \Psi_3 S = \begin{bmatrix} 0.009 & -0.002 \\ -0.002 & -0.013 \end{bmatrix}$$

⋮

The orthogonalised impulse response functions are then, for $y_1 = \Delta RS$ and $y_2 = \Delta R20$,

$$\psi_{12,0}^O = 0, \quad \psi_{12,1}^O = \begin{bmatrix} 1 & 0 \end{bmatrix} \begin{bmatrix} 0.131 & 0.071 \\ 0.035 & 0.078 \end{bmatrix} \begin{bmatrix} 0 \\ 1 \end{bmatrix} = 0.071, \quad \psi_{12,2}^O = 0.021, \ldots$$

$$\psi_{21,0}^O = 0.013, \quad \psi_{21,1}^O = \begin{bmatrix} 0 & 1 \end{bmatrix} \begin{bmatrix} 0.131 & 0.071 \\ 0.035 & 0.078 \end{bmatrix} \begin{bmatrix} 1 \\ 0 \end{bmatrix} = 0.035, \quad \psi_{21,2}^O = 0.001,$$

These response functions, along with their accumulations, are shown in Figure 7.1. Also shown are their counterparts when the ordering is reversed. There is a considerable difference between the two, showing clearly how a sizeable contemporaneous correlation between the innovations can alter the impulse responses. Nevertheless, the response of ΔRS to an innovation in $\Delta R20$ is clearly complete within six months and there is a smooth convergence of the accumulated response to a new positive 'level'. The response of $\Delta R20$ to an ΔRS innovation is very small when $\Delta R20$ is ordered first.

Figure 7.2 shows the generalised impulse response functions. The generalised responses for ΔRS are similar to the orthogonalised responses when $\Delta R20$ is first in the ordering and vice versa for $\Delta R20$ itself. Figure 7.3 shows the associated variance decompositions when $\Delta R20$ is first in the ordering. These show that innovations to $\Delta R20$ explain around 25% of the variation in ΔRS but that innovations to ΔRS explain none of the variation in $\Delta R20$.

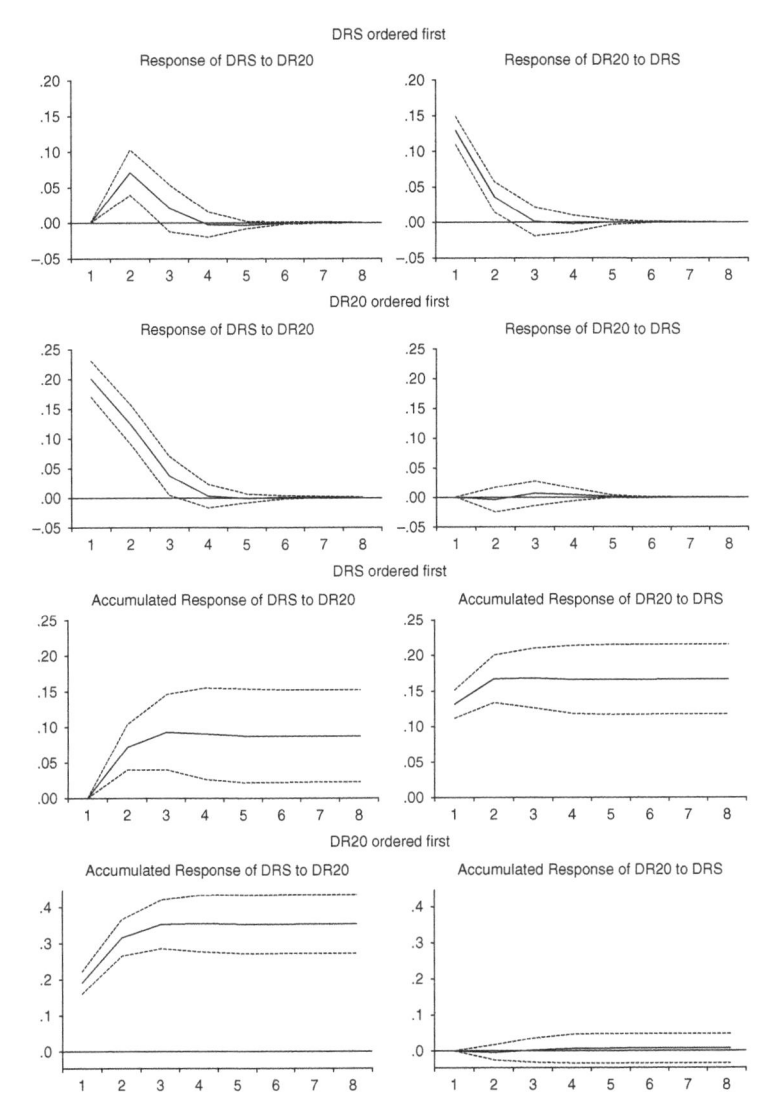

Figure 7.1 Orthogonalised impulse response functions: DRS and DR20 denote ΔRS and $\Delta R20$ respectively; 2 standard error bounds shown as dashed lines

110   *Time Series Econometrics*

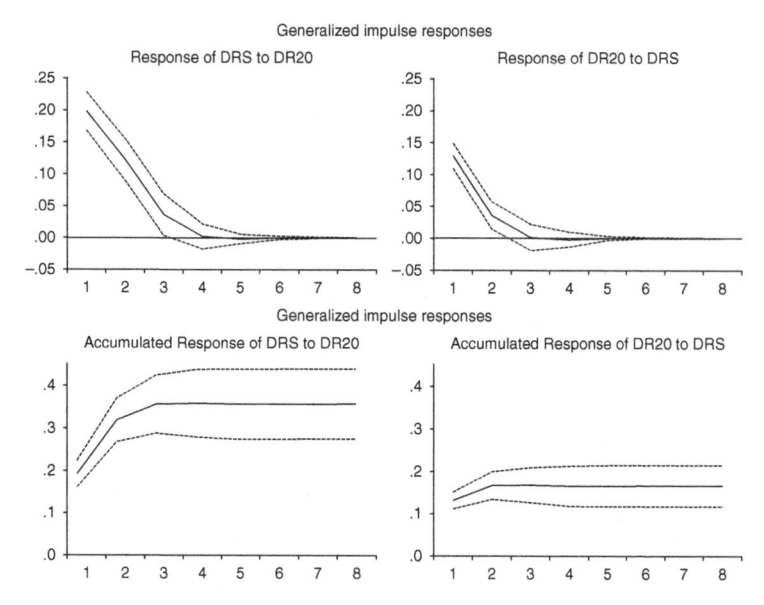

Figure 7.2   Generalised impulse response functions

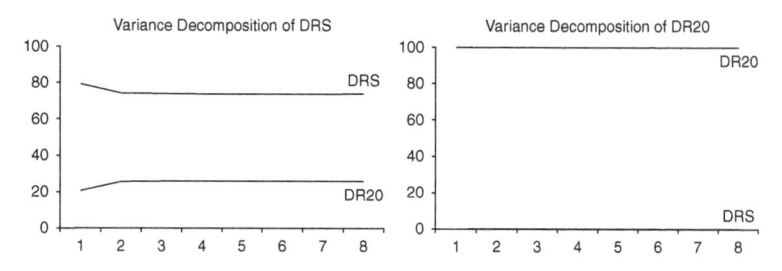

Figure 7.3   Variance decompositions

EViews Exercises

7.16   To estimate and analyse the VARs of Examples 7.1–7.3, open page `Ex _ 2 _ 2` and first explicitly generate the first differences of short and long interest rates:

```
genr drs = d(rs)
genr dr20 = d(r20)
```

Opening these variables as a group, clicking *Open Var* ... and clicking OK will, by default, estimate a VAR(2) with a constant vector included. To obtain the lag order statistics of Table 7.1, click *View/ Lag structure/Lag Length Criteria* and choose 4 as the 'Lags to include'. To estimate a VAR(1) without the (insignificant) constant vector, click *Estimate*, change 'Lag intervals for Endogenous' to 1 1 and remove c from the list of 'Exogenous Variables'. Clicking *View/ Residual Tests/Correlograms* will produce plots of the residual autocorrelation functions and cross-correlation functions for the default setting of 12 lags. Note the significant negative residual autocorrelation at lag 2 in the dr20 equation. Repeating the estimation with 'Lag intervals for Endogenous' changed back to 1 2 will estimate the VAR(2) without c: the residual correlograms now indicate no misspecification.

7.17 The Granger-causality test statistics reported in Example 7.2 may then be obtained by clicking *View/Lag Structure/Granger Causality-Block Exogeneity Tests*.

Plots of impulse response functions may be obtained by clicking *Impulse* and OK. The default setting is to use a Cholesky decomposition with the order of the variables being that of the original group selection, which in this case should be drs dr20; this ordering can be changed on clicking *Impulse definition* and it is here where other types of impulses, such as generalised impulses, may be selected. The number of periods over which the impulses are computed and whether the impulses are accumulated may also be chosen in the *Display* window.[11]

To compute variance decompositions click *View/Variance Decomposition* and complete the boxes of the view accordingly.

Notes

1. This model is treated in some detail in Aris Spanos, *Statistical Foundations of Econometric Modelling* (Cambridge University Press, 1986; Chapter 24).
2. The concepts of endogeneity and exogeneity being used here are the simplest possible: essentially a variable is termed endogenous if it is determined within the model, exogenous if it is determined outside of the model. These terms are deliberately kept loose but there are various tighter definitions in use for the models being discussed here. For an introductory text book discussion of these concepts, see Mills and Markellos, *Econometric Modelling*, chapter 8.6; for more detailed treatment, see Hendry, *Dynamic Econometrics* (Oxford University Press, 1995).

3. When the lag lengths p and q are the same across all equations then each of the equations will contain the same regressors. The model is then of a special type that can be efficiently estimated by OLS applied to each equation separately, known as multivariate least squares. When the lag lengths differ across equations this result no longer holds and a systems estimator must be used. A natural estimator is then **seemingly unrelated least squares**: see Arnold Zellner, 'An efficient method of estimating seemingly unrelated regressions and tests of aggregation bias', *Journal of the American Statistical Association* 57 (1962), 348–68.
4. The VAR was brought to the attention of economists by Christopher A. Sims, 'Macroeconomics and reality', *Econometrica* 48 (1980), 1–48, although a more general model, the vector ARMA, had been introduced over 20 years earlier by Maurice H. Quenouille, *The Analysis of Multiple Time Series* (Griffin, 1957). Even more general is the **dynamic structural equation model** (DSEM), which extends the multivariate dynamic regression model in two directions; first, by allowing simultaneity between the endogenous variables and, second, by explicitly considering the process generating the exogenous variables, for example

$$A_0 y_t = c + \sum_{i=1}^{m} A_i y_{t-i} + \sum_{i=1}^{m} B_i x_{t-i} + u_{1t}$$

$$x_t = d + \sum_{i=1}^{m} C_i x_{t-i} + u_{2t}$$

where $A_0 \neq I_n$. If there are no exogenous variables then the model reduces to a *structural VAR* (SVAR). For a development of these models, see Helmut Lütkepohl, *New Introduction to Multiple Time Series* (Springer Verlag, 2005).
5. The seminal papers on causality are Granger, 'Investigating causal relations by econometric models and cross-spectral methods', *Econometrica* 37 (1969), 424–38, and Sims, 'Money, income and causality', *American Economic Review* 62 (1972), 540–52. Although Granger provided an illustrative example to show the potential usefulness of the concept, he couched causality in a cross-spectral framework (which is generally unappealing to many economists) in which an estimation and testing methodology was not developed. Thus an appreciation of the concept's importance had to wait until a time domain approach to estimation and testing was developed, and this was provided soon after by Sims, who certainly helped to further popularise the concept by choosing as an example the then very 'hot' topic of the causal links between money and income.

Granger fully recognised that a precursor of his causality framework had been proposed over a decade earlier by Norbert Wiener ('The theory of prediction', in E.F. Breckenback (editor), *Modern Mathematics for Engineers* (McGraw-Hill, 1956), 165–90) and he typically referred to it as Wiener–Granger causality. For a detailed treatment of the concept, see Mills, *A Very British Affair*, chapter 9.
6. See, for example, Mills, *Analysing Economic Data*, §5.4.

7. See, for example, Mills, *Analysing Economic Data*, §13.3.
8. See Sims, 'An autoregressive index model for the US 1948–1975', in Jan Kmenta and James B. Ramsey (editors), *Large-scale Macroeconometric Models* (North-Holland, 1981), 283–327.
9. This 'non-invariance property' has generated much detailed analysis and criticism of the variance decomposition methodology, mainly focusing on the inability of VARs to be regarded as 'structural' in the traditional econometric sense, so that shocks cannot be uniquely identified with a particular variable unless prior identifying assumptions are made, without which the computed impulse response functions and variance decompositions would be invalid. The triangular 'recursive' structure of S has been criticised for being **atheoretical** and has led to the development of other sets of identifying restrictions that are based more explicitly on economic considerations using the SVAR approach: see Thomas F. Cooley and Stephen F. LeRoy, 'Atheoretical macroeconometrics: a critique', *Journal of Monetary Economics* 16 (1985), 283–308, and Olivier J. Blanchard, 'A traditional interpretation of macroeconomic fluctuations', *American Economic Review* 79 (1989), 1146–64.
10. These were introduced by M. Hashem Pesaran and Yeongcheol Shin, 'Generalized impulse response analysis in linear multivariate models', *Economics Letters* 58 (1997), 17–29.
11. The vector MA coefficient matrices can be read off from the impulse response function when 'Residual-one unit' is selected as the Decomposition Method and 'Table' is selected as the Display Format. The first (second) row of ψ_j is given by the response of drs (dr20) in period $j + 1$.

Cointegration in Single Equations

8

Cointegration in Single Equations

Spurious regression

8.1 The VAR framework of the previous chapter requires that all the time series contained in the model be stationary. Whilst stationarity can be achieved, if necessary, by differencing each of the individual series, is this always an appropriate approach to take when working within an explicitly multivariate framework? We begin our answer to this question by introducing the simulation example considered by Clive Granger and Paul Newbold in an important article examining some of the likely empirical consequences of nonsense, or *spurious*, regressions in econometrics.[1]

Granger and Newbold proposed a simulation set-up in which y_t and x_t are generated by the *independent* random walks

$$y_t = y_{t-1} + v_t, \quad x_t = x_{t-1} + w_t, \quad t = 1, 2, \ldots, \tag{8.1}$$

where v_t and w_t are independent white noises. The regression of y_t on a constant and x_t is then considered:

$$y_t = \hat{\alpha}_T + \hat{\beta}_T x_t + \hat{u}_t, \quad t = 1, 2, \ldots, T \tag{8.2}$$

With $T = 50$, $y_0 = x_0 = 100$ and v_t and w_t drawn from independent $N(0,1)$ distributions, the regression (8.2) was estimated for 100 pairs of y_t and x_t series generated using (8.1). Clearly, since v_t and w_t are independent white noises, y_t and x_t must themselves be independent integrated processes and so there should be no relationship between them. However, for these 100 regressions, Granger and Newbold reported that in 76 of them the (correct) null hypothesis of $\beta = 0$ was rejected using the conventional t-statistic for assessing the significance of $\hat{\beta}_{50}$ at the 5% level (so the critical value is approximately 2). In other words, rather than a rejection rate of 5%, they observed a rejection rate of 76%! In fact, they showed that to ensure a 5% rejection rate a critical t-value of over 10 should be used, as the standard deviation of $\hat{\beta}_{50}$ was being under-estimated by a factor of over 5.[2]

8.2 When five independent random walks were included as regressors in a multiple regression, things got even worse, for the rejection rate of a conventional F-statistic testing that the entire coefficient vector is zero rose to 96%. For regressions involving independent ARIMA(0,1,1) series the corresponding rejection rates were 64% and 90%. Granger and Newbold thus concluded that, when regressors were generated as statistically independent integrated processes, conventional significance tests were seriously biased towards rejection of the null hypothesis of no relationship, and hence towards the acceptance of a *spurious* relationship.

8.3 Moreover, they also found that the regressions were frequently accompanied by large R^2 values and highly autocorrelated residuals, as indicated by very low Durbin-Watson (dw) statistics. These findings led Granger and Newbold to suggest that, in the joint circumstances of a high R^2 and a low dw statistic (a useful rule being $R^2 > dw$), regressions should be run on the first differences of the variables, so providing support for the practice of differencing to induce stationarity in time series regressions.

8.4 These essentially empirical conclusions were later given an analytical foundation by Peter Phillips, who showed that the standard distributional results of least squares regression actually broke down when regressors were integrated processes.[3] Phillips obtained four analytical results that went a long way towards explaining the simulation findings reported by Granger and Newbold, although it suffices here to provide purely verbal and graphical explanations.

The first result shows that, in contrast to standard regression theory, $\hat{\alpha}_T$ and $\hat{\beta}_T$ do not converge in probability to constants as

$T \to \infty$. $\hat{\beta}_T$ has, in fact, a non-degenerate limiting distribution (the sampling distribution does not converge to the single true value of β as $T \to \infty$), so that different, yet arbitrary, large samples will yield randomly differing estimates of β. The distribution of $\hat{\alpha}_T$ actually diverges, so that estimates are likely to get farther and farther away from the true value of α as the sample size increases. Thus the uncertainty about the regression (8.2) stemming from its spurious nature is not a small sample problem, for it persists asymptotically in these limiting distributions.

The second result shows that the conventional t-ratios on $\hat{\alpha}_T$ and $\hat{\beta}_T$ do not have t-distributions, and indeed do not have *any* limiting distribution, diverging as $T \to \infty$ so that there are *no* asymptotically correct values for these tests. We should thus expect the rejection rate when tests are based on a critical value delivered from conventional asymptotics (such as 1.96) to continue to increase with sample size, and this therefore explains the findings of Granger and Newbold.

The third and fourth results show that R^2 has a non-degenerate limiting distribution and that dw converges in probability to 0 as $T \to \infty$. Low values for dw and moderate values of R^2 are therefore to be expected in spurious regressions such as (8.2) with data generated by integrated processes, again confirming the simulation findings reported by Granger and Newbold.

8.5 These results are easily extended to multiple regressions of the form

$$y_t = \hat{\alpha}_T + \boldsymbol{\beta}'_T \mathbf{x}_t + \hat{u}_t \tag{8.3}$$

where $\mathbf{x}_t = (x_{1t},\ldots,x_{kt})'$ is a vector of $I(1)$ processes. Phillips showed that analogous results to those discussed in §8.4 hold for (8.3) and, in particular, that the distribution of the customary F-statistic for testing a set of linear restrictions on $\boldsymbol{\beta}$ diverges as $T \to \infty$ and so there are no asymptotically correct critical values for this statistic either. Moreover, the divergence rate for the F-statistic is greater than that for individual t-tests, implying that, in a regression with many regressors, we might expect a noticeably greater rejection rate for a 'block' F-test than for individual t-tests or for a test with fewer regressors, and this is again consistent with the results reported by Granger and Newbold.

8.6 We should emphasise that, although y_t and x_t have been assumed to be independent in (8.2) and (8.3), so that the true values of α and β are zero, this is not crucial to the major conclusions. Although the correlation properties of the time series do have quantitative effects on the limiting distributions, such effects do not interfere with the main qualitative results. To reiterate, these are that $\hat{\alpha}_T$ and $\hat{\beta}_T$ do not converge in probability to constants, the distributions of F- and t-statistics diverge as $T \to \infty$, dw converges in probability to 0, and R^2 has a non-degenerate limiting distribution as $T \to \infty$.

8.7 A simulation similar to that of Granger and Newbold enables us to interpret these results in a perhaps more transparent fashion. The independent random walks y_t and x_t were generated for a sample now of size $T = 1000$, with v_t and w_t again drawn from independent $N(0,1)$ populations with $y_0 = x_0 = 0$, using 10,000 iterations.

Figures 8.1 to 8.4 present the density functions of $\hat{\beta}_{1000}$, its associated t-ratio, and the R^2 and dw statistics. The distribution of $\hat{\beta}_{1000}$ shown in Figure 8.1 is almost normally distributed (a central limit theorem does, in fact, hold as the simulations use independent replications). Although the sample mean of -0.005 is very close to the true

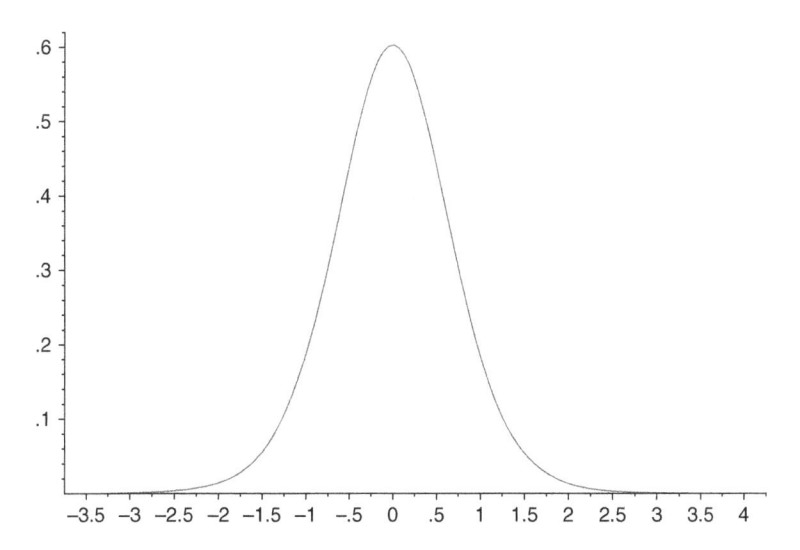

Figure 8.1 Simulated frequency distribution of $\hat{\beta}_{1000}$

β value of 0, the sample standard deviation is 0.635, thus confirming that, for large T, the distribution does not converge to a constant and different samples produce very different estimates of β, the range of estimates being approximately ±3.0.

The distribution of the t-ratio, shown in Figure 8.2, is again normal but with a standard deviation of 23.62. The 5% critical values from this distribution are ±48.3 while using a value of ±1.96 would entail a rejection rate of 93.4%! The distribution of the R^2 statistic, shown in Figure 8.3, has a mean of 0.24, a standard deviation of 0.23 and a maximum value of 0.94, while that for dw (Figure 8.4) has a mean of 0.018, a standard deviation of 0.011, and a maximum value of only 0.10.[4] Both sampling distributions thus illustrate the theoretical predictions of Phillips' analysis.

8.8 It should be emphasised that, in the general spurious regression set-up discussed here, both y_t and x_t are $I(1)$ processes. The error, u_t, since it is by definition a linear combination of $I(1)$ processes, will therefore also be integrated, unless a special restriction, to be discussed in §8.12, holds. Moreover, the usual re-specification of the model to include y_{t-1} as an additional regressor on the finding of a very low dw value will have pronounced consequences. The

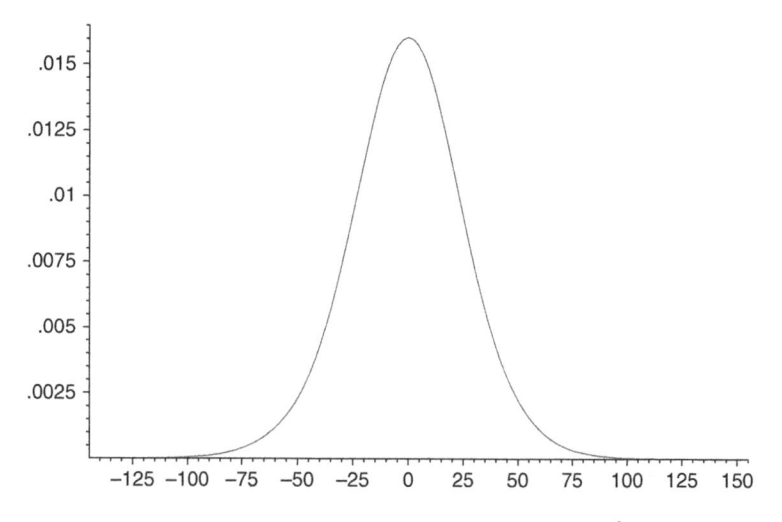

Figure 8.2 Simulated frequency distribution of the t-ratio of $\hat{\beta}_{1000}$

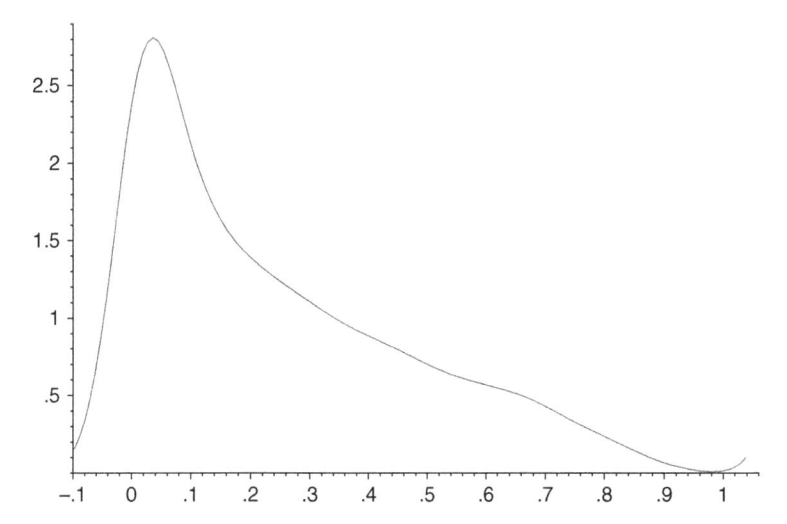

Figure 8.3 Simulated frequency distribution of the spurious regression R^2

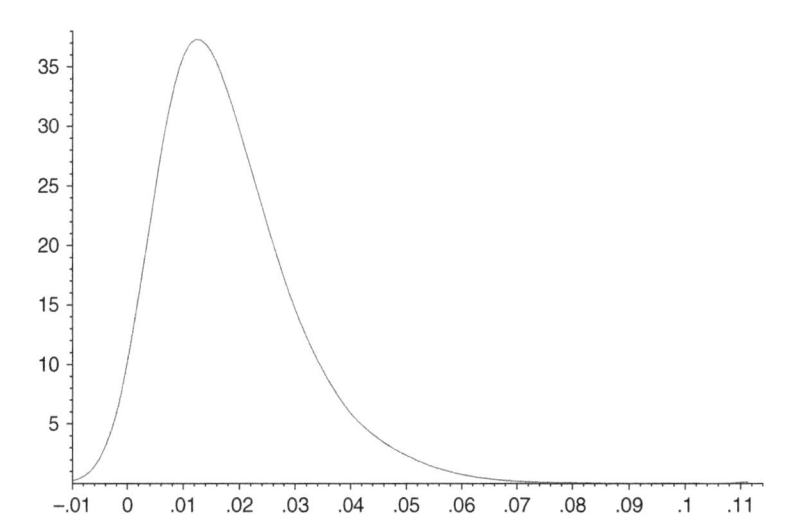

Figure 8.4 Simulated frequency distribution of the spurious regression *dw*

estimated coefficient on y_{t-1} will converge to unity, while those on the integrated regressors will converge to 0, thus highlighting the spurious nature of the static regression. Indeed, the spurious nature of the regression is, in fact, a consequence of the error being $I(1)$. Achieving a stationary, or $I(0)$, error is usually a minimum criterion to meet in econometric modelling, for much of the focus of recent developments in the construction of dynamic regression models has been to ensure that the error is not only $I(0)$ but white noise, and this underlies the general-to-specific model building philosophy of dynamic econometrics.[5] Whether the error in a regression between integrated variables is stationary is thus a matter of considerable empirical importance.

8.9 It is fair to say that many econometricians were extremely sceptical of Granger and Newbold's 'prescription' of differencing variables to induce stationary regressors. This scepticism stemmed from noting that the static regression $y_t = \alpha + \beta x_t + u_t$ with errors following the *stationary* process $u_t = \rho u_{t-1} + \varepsilon_t$, $|\rho| < 1$, can be written as

$$y_t = \alpha(1-\rho) + \beta x_t - \rho\beta x_{t-1} + \rho y_{t-1} + \varepsilon_t \qquad (8.4)$$

which is a restricted version of the general dynamic regression

$$y_t = \gamma_0 + \gamma_1 x_t + \gamma_2 x_{t-1} + \gamma_3 y_{t-1} + w_t \qquad (8.5)$$

The restriction imposed on (8.5) to obtain (8.4) is $\gamma_2 + \gamma_1\gamma_3 = 0$ and whether this restriction is satisfied or not determines how regressions with autocorrelated variables should best be modelled: if the restriction is satisfied then the relationship between y_t and x_t is appropriately modelled as a static regression with an autocorrelated error; if it is not then the correct specification is the dynamic regression (8.5).[6]

8.10 This set-up still lies within a stationary environment but suppose (8.5) is now considered in isolation. Imposing the pair of restrictions $\gamma_1 + \gamma_2 = 0$ and $\gamma_2 = 1$ on (8.5) produces the 'restriction form'

$$\Delta y_t = \gamma_0 + \gamma_1 \Delta x_t + w_t \qquad (8.6)$$

This may be compared with the 'operator form', implicitly considered by Granger and Newbold and which transforms (8.5) by differencing throughout to obtain

$$\Delta y_t = \gamma_1 \Delta x_t + \gamma_2 \Delta x_{t-1} + \gamma_3 \Delta y_{t-1} + \Delta w_t \qquad (8.7)$$

If the coefficient restrictions imposed on (8.5) to obtain (8.6) are valid then, if w_t is white noise in (8.5) (or, in general, stationary), so must be the error in (8.6). Moreover, the operator form (8.7) is then misspecified in that it incorrectly includes Δx_{t-1} and Δy_{t-1}, excludes the intercept, and has a non-invertible moving average error. Alternatively, if w_t is a random walk in (8.5) (more generally, $I(1)$) then the operator form (8.7) becomes the correct specification.[7]

8.11 Even so, both (8.6) and (8.7) have some unacceptable features in terms of being universally valid formulations for economic systems. In particular, (8.6) either has *no* equilibrium solution in terms of y_t and x_t or one that collapses to zero if $\gamma_1 = 0$. Moreover, in either specification the time paths that y_t can describe will be independent of the states of disequilibrium existing in prior periods.

From this perspective, there must be other ways of transforming to stationarity than by just differencing, with the choice of which transformation to adopt potentially being based on considerations from economic theory. For example, while marginal adjustments might favour differencing, long-run considerations could suggest specifications of the form

$$\Delta y_t = \gamma_0 + \gamma_1 \Delta x_t + (\gamma_2 - 1)(y_{t-1} - x_{t-1}) + w_t \qquad (8.8)$$

which is obtained from (8.5) by imposing the restriction that $\gamma_1 + \gamma_2 + \gamma_3 = 1$. This allows an equilibrium solution to emerge when $\Delta y_t = \Delta x_t = w_t = 0$, for then

$$y_t = x_t + \frac{\gamma_0}{1 - \gamma_2}$$

which would be appropriate if y_t and x_t are logarithms and there is postulated to be a long-run unit elasticity between the variables.

Equations of the form (8.8) quickly became known as *error correction models*, as the term $y_{t-1} - x_{t-1}$ measures the extent to which the system is away from equilibrium and hence it represents the error that must be 'corrected'.[8]

Cointegrated processes

8.12 Notwithstanding the discussion of §§8.9-8.11, the general implication of the above analysis is that a linear combination of $I(1)$ processes will usually also be $I(1)$. In general, if y_t and x_t are both $I(d)$, then the linear combination

$$u_t = y_t - ax_t \tag{8.9}$$

will also usually be $I(d)$. It is possible, however, that u_t may be integrated of a lower order, say $I(d-b)$, where $b > 0$, in which case a special constraint operates on the long-run components of the two series.[9]

If $d = b = 1$, so that y_t and x_t are both $I(1)$ and dominated by long-run components (recall Figure 3.3), u_t will be $I(0)$ and hence will not contain any such components: y_t and ax_t must therefore have long-run components that cancel out to produce a stationary u_t. In these circumstances, y_t and x_t are said to be *cointegrated*; we emphasise that it will *not* generally be true that there will exist an a which makes $u_t \sim I(0)$ or, in general, $I(d-b)$.[10]

8.13 The idea of cointegration can be related to the concept of *long-run equilibrium* discussed in §8.11, a connection which we may illustrate with the bivariate relationship

$$y_t = ax_t$$

or

$$y_t - ax_t = 0$$

In this context, u_t given by (8.9) measures the extent to which the 'system' is out of equilibrium, and it can therefore be termed the *equilibrium error*. Assuming that $d = b = 1$, so that y_t and x_t are both

$I(1)$, the equilibrium error will then be $I(0)$ and u_t will rarely drift far from 0, and will often cross the zero line. In other words, equilibrium will occasionally occur, at least to a close approximation, whereas if y_t and x_t are not cointegrated, so that $u_t \sim I(1)$, the equilibrium error will wander widely and zero-crossings would be rare, suggesting that under these circumstances the concept of equilibrium has no practical implications (recall the distinctions between the properties of $I(0)$ and $I(1)$ processes listed in §4.1).

8.14 How, though, is the concept of cointegration linked to the analysis of spurious regressions? To answer this, we need to define the covariance matrix

$$\Sigma_s = \begin{bmatrix} \sigma_v^2 & \sigma_{vw} \\ \sigma_{vw} & \sigma_w^2 \end{bmatrix}$$

where σ_{vw} is the covariance between v_t and w_t, which is not necessarily 0. The results of §8.4 actually require Σ_s to be non-singular. If this is not the case, then the asymptotic theory yielding these results no longer holds.

For Σ_s to be singular, we require $|\Sigma_s| = \sigma_v^2 \sigma_w^2 - \sigma_{vw}^2 = 0$. This implies that $\Sigma_s \gamma = 0$, where $\gamma' = (1, -a)$ and $a = \sigma_{vw}/\sigma_w^2$. The singularity of Σ_s is, in fact, a necessary condition for y_t and x_t to be cointegrated, since in this case $|\Sigma_s| = 0$ implies that the correlation between the innovations v_t and w_t, given by $\rho_{vw} = \sigma_{vw}/\sigma_v \sigma_w$, is unity. For values of ρ_{vw} less than unity, y_t and x_t are not cointegrated, although they are clearly correlated, and when $\rho_{vw} = 0$, so that v_t and w_t are independent, we have Granger and Newbold's spurious regression.

8.15 What differences to the asymptotic regression theory for integrated regressors outlined in §§8.4–8.7 are there when y_t is cointegrated with x_t? Since the equilibrium error u_t can be regarded as the error term in the regression of y_t on x_t, we may consider the model

$$y_t = \beta x_t + u_t \quad (8.10)$$

$$x_t = x_{t-1} + w_t$$

where u_t and w_t are contemporaneously correlated white noise, so that $E(u_t w_t) = \sigma_{uw}$. This non-zero correlation implies that x_t is endogenous rather than exogenous.

Several theoretical results can be demonstrated via simulation. The model given by (8.10) was used with the correlation between u_t and w_t parameterised as $u_t = \gamma w_t + v_t$, so that $\gamma = \sigma_{vw}/\sigma_w^2$. The simulations set $\beta = 0$, $\sigma_w^2 = \sigma_u^2 = 1$ and $\sigma_{uw} = 0.75$, so that $\gamma = 0.75$. With once again $T = 1000$ and 10,000 iterations, Figure 8.5 shows the simulated frequency distribution of $\hat{\beta}_{1000}$. The sample mean is 0.0028 and 95% of the estimates lie in a very small interval, −0.0016 to 0.0093, around 0, reflecting what is known as the *super-consistency* property of cointegration. However, this interval also shows the skewness of the distribution, which is a consequence of the presence of endogeneity bias caused by the lack of exogeneity of x_t (this is known as second-order bias).

Figure 8.6 shows the simulated t-ratio. The distribution is not standard normal, although it is normal in shape, being centred not on 0 but on 0.994 and with a standard deviation of just 0.884, so that the 2.5% critical value is approximately 2.8 rather than 2.

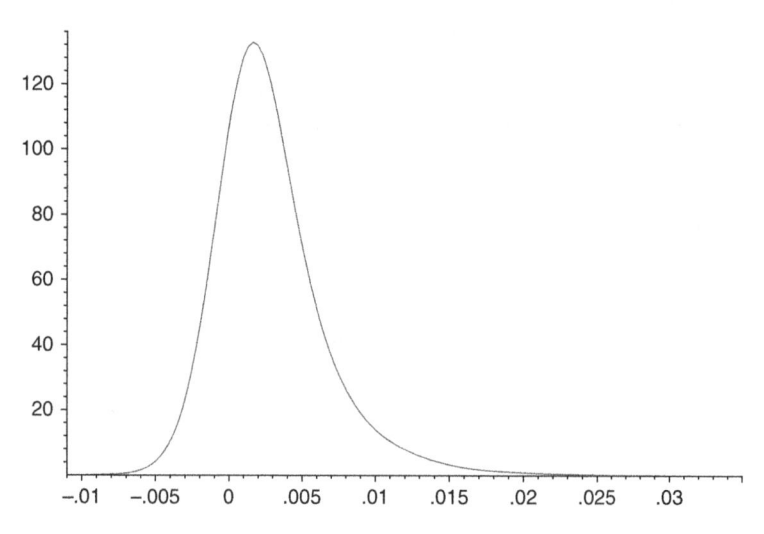

Figure 8.5 Simulated frequency distribution of $\hat{\beta}_{1000}$ from the cointegrated model with endogenous regressor

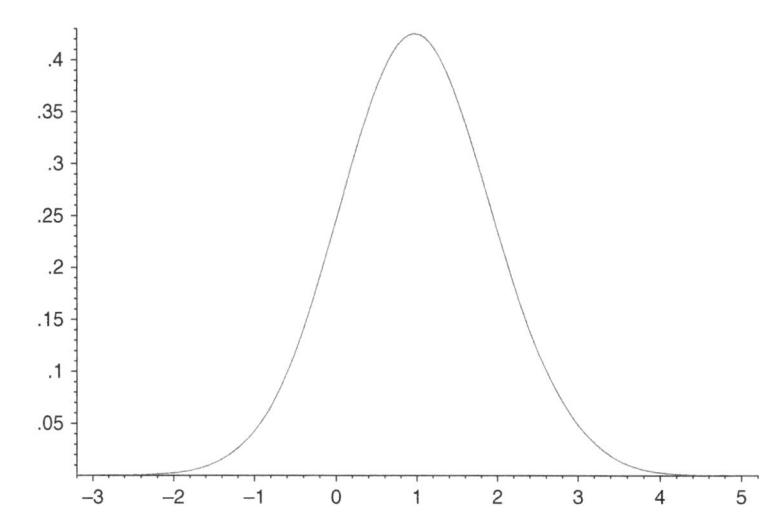

Figure 8.6 Simulated frequency distribution of the t-ratio on $\hat{\beta}_{1000}$ from the cointegrated model with endogenous regressor

8.16 Figures 8.7–8.9 show the results of three related simulations. Figure 8.7 provides the simulated frequency distribution of the slope coefficient of the regression of y_t on x_t when x_t is generated by the stationary AR(1) process $x_t = 0.5x_{t-1} + w_t$ rather than the random walk of (8.10), but where all other settings remain the same. The endogeneity bias is now readily apparent, with the distribution, although normal, having a mean of 0.565 rather than 0, and a standard deviation of 0.035.

Figure 8.8 shows the simulated frequency distribution of the slope coefficient in the same stationary regression but where now $\sigma_{uw} = 0$, so that there is no endogeneity: consequently, the distribution is centred on 0. Finally, Figure 8.9 shows the frequency distribution of $\hat{\beta}_{1000}$ from the cointegrated model but with $\sigma_{uw} = 0$. With no endogeneity, the distribution is normal, as compared to Figure 8.5, but has a standard error of just 0.0035, thus reflecting the super-consistency property of a cointegrated regression when compared to its stationary counterpart in Figure 8.8.

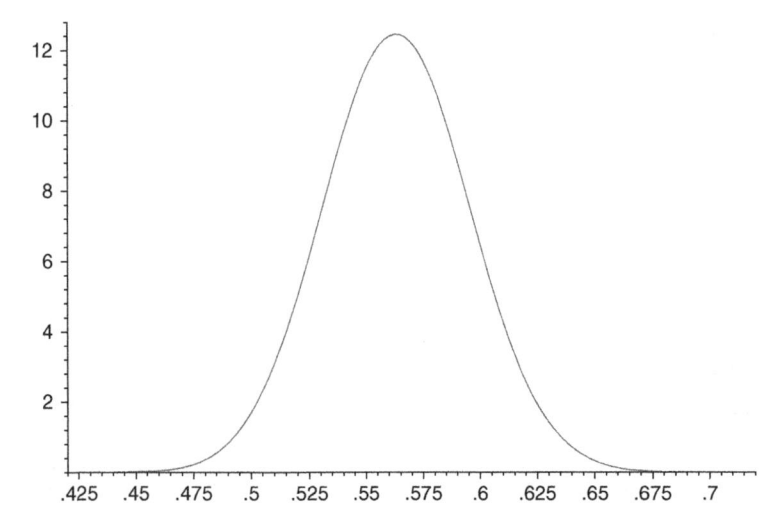

Figure 8.7 Simulated frequency distribution of the slope coefficient from the stationary model with endogeneity

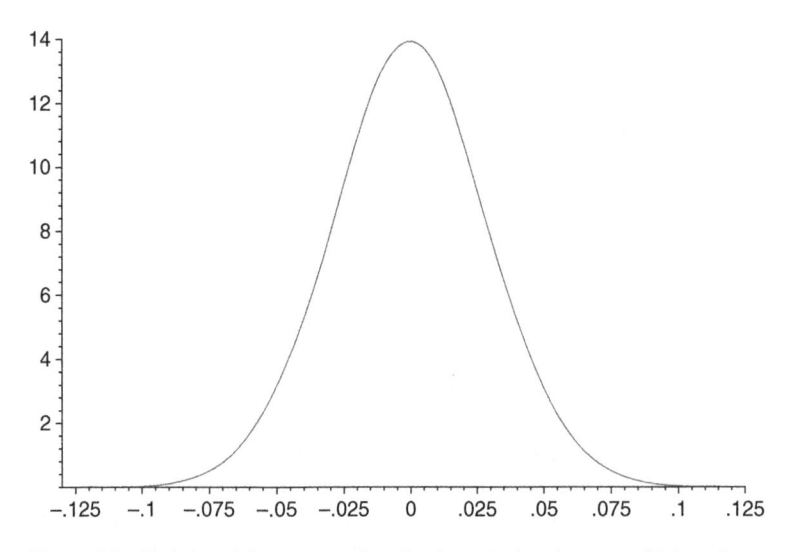

Figure 8.8 Simulated frequency distribution of the slope coefficient from the stationary model without endogeneity

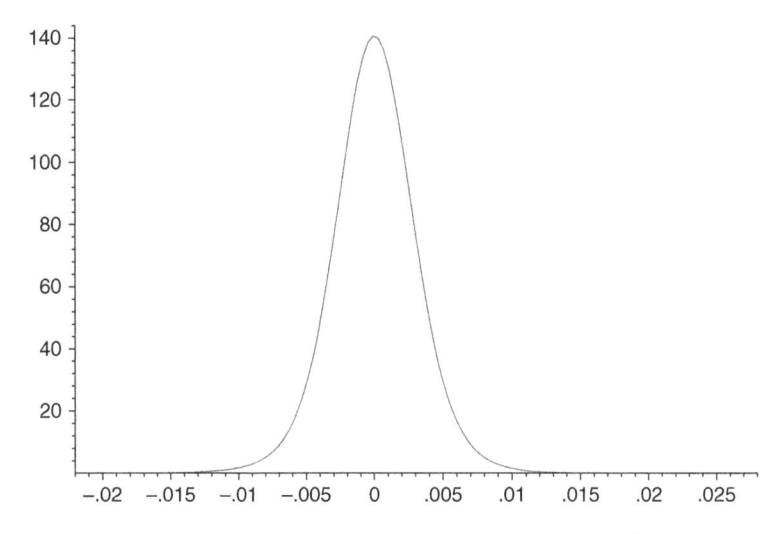

Figure 8.9 Simulated frequency distribution of the t-ratio on $\hat{\beta}_{1000}$ from the cointegrated model with exogenous regressor

8.17 The assumption made in all these simulations is that x_t is without drift. This assumption is not innocuous, however, for the inclusion of a drift restores the asymptotic normality of $\hat{\beta}_T$ irrespective of whether there is endogeneity or not. For multiple regressions this result does not hold although the estimator continues to be super-consistent.[11]

Testing for cointegration in regression

8.18 Given the crucial role that cointegration plays in regression models with integrated variables, it is clearly important to be able to test for its presence. A number of tests have been proposed that are based on the residuals from the *cointegrating regression*

$$\hat{u}_t = y_t - \hat{\alpha}_T - \hat{\boldsymbol{\beta}}_T \mathbf{x}_t \tag{8.11}$$

Such residual-based procedures seek to test a null hypothesis of *no* cointegration. Perhaps the simplest test to use is the usual

Durbin–Watson dw statistic but, since the non-cointegration null is that $\hat{u}_t$ is $I(1)$, the value of the test statistic under this null is $dw = 0$, with rejection in favour of the $I(0)$ alternative occurring for values of dw greater than 0. Unfortunately, there are several difficulties associated with this simple test which mirror those that affect the conventional Durbin–Watson test. For example, the asymptotic distribution of dw under the null depends on nuisance parameters such as the correlations within the vector $\Delta \mathbf{x}_t$; the critical value bounds diverge as the number of regressors increases, becoming so wide as to have no practical value for inference; and the statistic assumes that under the null u_t is a pure random walk, and under the alternative u_t is a stationary AR(1) process. If this actually is the case, then dw has excellent power properties, but the critical bounds will not be correct if u_t has more complicated autocorrelation properties.[12]

8.19 A more popular test is to use the t-ratio on $\hat{u}_{t-1}$ from the regression of $\Delta\hat{u}_t$ on $\hat{u}_{t-1}$ and lagged values of $\Delta\hat{u}_t$, in a manner analogous to the unit root testing approach for an observed series discussed in §§4.4–4.8. This is often known as the Engle–Granger test for cointegration. The problem here is that $\Delta\hat{u}_t$ is derived as a residual from a regression in which the cointegrating vector is estimated. If the null of non-cointegration was actually true then such a vector would not be identified. However, least squares will nevertheless estimate the cointegrating vector which minimises the residual variance and hence is most likely to result in a stationary residual series, so that using the Dickey–Fuller τ_μ critical values would reject the null too often. Moreover, an additional factor that influences the distribution of the t-ratio is the number of regressors contained in $\mathbf{x}_t$. Critical values are available from many sources: for example, the large T 5%, 2.5% and 1% critical values when $\mathbf{x}_t = x_t$ are –3.37, –3.64 and –3.96. As with conventional unit root tests, different sets of critical values are to be used if there is either no constant in the cointegrating regression or if there is both a constant and a trend (corresponding to the τ and τ_τ variants).

EXAMPLE 8.1 Are short and long interest rates cointegrated?

The findings that the interest rate spread, the difference $R20_t - RS_t$, is $I(0)$ in Example 4.1, and that individually RS_t and $R20_t$ are both

$I(1)$ in Example 4.4, may be interpreted as implying that RS_t and $R20_t$ are cointegrated. Note that the plot of the two series in Figure 4.5 certainly suggests that they are 'bound together' through time. The Engle–Granger (EG) test for cointegration is obtained by first regressing RS_t on a constant and $R20_t$ and then subjecting the residuals to a unit root test. Doing so yields $\tau_\mu = -3.94$, which is significant at the 1% level (the marginal significance level being 0.009). Of course, choosing RS_t to be the dependent variable in the cointegrating regression is arbitrary: we could just as well regress $R20_t$ on a constant and RS_t. If we do this we obtain $\tau_\mu = -3.48$, which is significant at 5% (marginal significance level 0.036). We are thus able to confirm that RS_t and $R20_t$ are indeed cointegrated.

Estimating cointegrating regressions

8.20 As we have seen, OLS estimation of the cointegrating regression produces estimates that, although super-consistent, are nevertheless (second-order) biased even in large samples (recall Figure 8.5, which showed a biased sampling distribution of $\hat{\beta}_{1000}$ when there was endogeneity between y_t and x_t: autocorrelation in u_t will exacerbate the situation further).

A general set-up that allows for both contemporaneous correlation and autocorrelation is an extension of (8.10) to the multivariate 'triangular' system

$$y_t = \beta x_t + u_t \tag{8.12}$$

$$\Delta x_t = w_t$$

We can assume that u_t and w_t are stationary, but not necessarily white noise, processes which may be correlated. As discussed above, OLS estimates are second-order biased and, of course, this arises because of the contemporaneous and serial correlation of the regressors. This bias may be eliminated by using the **fully-modified OLS** (FM-OLS) estimator.[13]

8.21 Several other estimators have been proposed that correct for both correlation between u_t and w_t and autocorrelation in u_t while still continuing to use standard least squares techniques. The most popular of these is to augment (8.12) with leads and lags of $\Delta \mathbf{x}_t$ when there is correlation between u_t and $\mathbf{w}_t$:

$$y_t = \beta \mathbf{x}_t + \sum_{s=-p}^{p} \gamma_s \Delta \mathbf{x}_{t-s} + u \quad (8.13)$$

where p is chosen such that the correlation between u_t and $\mathbf{w}_t$ is zero for $|s| > p$. This is known as *dynamic OLS* (DOLS). If $\mathbf{x}_t$ is strongly exogenous, so that u_t does not Granger-cause $\mathbf{w}_t$, then leads of $\Delta \mathbf{x}_t$ will not be required ($\gamma_s = 0$, $s < 0$).[14]

Autocorrelation in u_t may be captured by assuming that u_t follows an AR(p) process and estimating (8.13) either by generalised least squares (GLS), by including lags of Δy_t as additional regressors, or by using a non-parametric correction of the standard errors.

EXAMPLE 8.2 Estimating the cointegrating regression between short and long interest rates

OLS estimation of the cointegrating regression $R20_t = \alpha + \beta RS_t + u_t$ produces

$R20_t = 2.536 + 0.788 \; RS_t + \hat{u}_t$
$(0.121) \; (0.016)$

but, as we have seen, such a regression should not be used for inference. FM-OLS estimation with a non-parametric correction for autocorrelation produces

$R20_t = 1.110 + 1.009 \; RS_t + \hat{u}_t$
$(0.973) \; (0.131)$

These estimates are very different to those from OLS and show that α is insignificantly different from 0 and β insignificantly different from unity; indeed, the joint hypothesis $\alpha = 0$ and $\beta = 1$ cannot be rejected using an F-test at the 5% level. The DOLS estimator with a similar correction for autocorrelation produces

$R20_t = 1.720 + 0.908 \; RS_t + \hat{u}_t$
$(0.336) \; (0.045)$

Cointegration in Single Equations 131

and is again consistent with the 'spread hypothesis' that $R20_t - RS_t$ is stationary, so that RS_t and $R20_t$ are cointegrated with *cointegrating vector* (1, −1). However, from Example 2.2, recall that the spread has a non-zero mean, which requires that α should be non-zero, in fact positive (the estimate was 1.128). The estimates of α from FM-OLS and DOLS are both consistent with this value.

EViews Exercises

8.22 In page Ex _ 2 _ 2 open r20 and rs as a group and click *View/ Cointegration Test/Single-Equation Cointegration Test...*. The default settings are appropriate and clicking OK produces the two EG test statistics (called the 'tau-statistics') reported in Example 8.1.

8.23 The standard OLS command

```
ls r20 c rs
```

produces the OLS estimated cointegrating regression of Example 8.2. To obtain the FM-OLS estimates click *Estimate* and select COINTREG – Cointegrating Regression as the estimation method. In the 'Specification' window click 'Options' in 'Nonstationary estimation settings' (noting that 'Fully-modified OLS (FMOLS)' is the default estimation method) and then choose 'Auto-AIC' as the lag specification, whereupon the estimates reported in Example 8.2 will be produced. The test of the joint hypothesis $\alpha = 0$ and $\beta = 1$ is obtained by clicking *View/Coefficient Diagnostics/Wald test – Coefficient Restrictions* in the 'Stats' window and entering c(1)=1,c(2)=0 in the coefficient restrictions box. The *F*-statistic is seen to be (just) insignificant at the 5% level.

The DOLS estimates may be obtained by clicking 'Estimate' and now selecting 'Dynamic OLS (DOLS)' as the method. Selecting 'Akaike' as the 'Lag & Lead method' will then produce the DOLS estimates.

Notes

1. Granger and Newbold, 'Spurious regressions in econometrics', *Journal of Econometrics* 2 (1974), 111–20. Spurious (or *nonsense*) regressions had first been analysed by Yule almost 50 years earlier in 'Why do we sometimes get nonsense-correlations between time-series? A study in sampling and the nature of time series', *Journal of the Royal Statistical Society* 89 (1926),

1–63. See Mills, *A Very British Affair*, chapters 2 and 10, for detailed discussion of the two papers.
2. Granger has recounted that when he presented these simulation results during a seminar presentation at the LSE they were 'met with total disbelief. Their reaction was that we must have done the programming incorrectly' (Peter C.B. Phillips, 'The ET interview: Professor Clive Granger', *Econometric Theory* 13 (1997), 262). Paul Newbold, who had actually done the programming, confirmed this story to me some years ago over a couple of pints in the Nottingham University Staff Club.
3. Phillips, 'Understanding spurious regressions in econometrics', *Journal of Econometrics* 33 (1986), 311–40.
4. Note that the smoothing involved in constructing the density functions leads to negative values in the left-hand tails of these two distributions: the actual minimum sample values of R^2 and dw are, of course, positive, although extremely small, being 0.0008 for dw and of the order of 10^{-10} for R^2.
5 See, for example, Hendry, *Dynamic Econometrics*.
6 These alternative specifications were emphasised in Hendry and Grayham E. Mizon, 'Serial correlation as a convenient simplification, not a nuisance: a comment on a study of the demand for money by the Bank of England', *Economic Journal* 88 (1978), 549–63.
7. These different interpretations and implications of differencing had, in fact, been discovered half a century earlier by Bradford B. Smith, 'Combining the advantages of first-difference and deviation-from-trend methods of correlating time series', *Journal of the American Statistical Association* 21 (1926), 55–9. This was a remarkably prescient article that quickly disappeared without trace until it was rediscovered by Mills, 'Bradford Smith: an econometrician decades ahead of his time', *Oxford Bulletin of Economics and Statistics* 73 (2011), 276–85.
8. These models were popularised by James Davidson, Hendry, Frank Srba and Stephen Yeo, 'Econometric modelling of the aggregate time-series relationship between consumers' expenditure and income in the United Kingdom', *Economic Journal* 88 (1978), 861–92.
9. Granger recalls discussing this possibility with Hendry: 'he was saying that he had a case where he had two $I(1)$ variables, but their difference was $I(0)$, and I said that is not possible, speaking as a theorist. He said he thought it was. So I went away to prove that I was right, and I managed to prove that he was right' (Phillips, 1997; page 25); also see Granger, 'Some thoughts on the development of cointegration', *Journal of Econometrics* 158 (2010), 3–6, for further recollections of this episode.
10. The concept of cointegration was first introduced in Granger, 'Some properties of time series data and their use in econometric model specification', *Journal of Econometrics* 16 (1981), 121–30, was further developed in Granger and Andrew A. Weiss, 'Time series analysis of error correcting models', in Samual Karlin, Takeshi Amemiya and Leo A. Goodman (editors), *Studies in Econometrics, Time Series and Multivariate Statistics*

(Academic Press, 1983), 255–78, and came to prominence with Engle and Granger, 'Cointegration and error correction: representation, estimation and testing', *Econometrica* 55 (1987), 251–76, which has since become the most heavily cited paper in time series econometrics. Such has been the impact of cointegration that Granger, jointly with Engle, was awarded the 2003 Severiges Riksbank Prize in Economic Science in memory of Alfred Nobel in 'recognition of his achievements in developing methods of analysing economic time series with common trends (co-integration)'. See Mills, *A Very British Affair*, chapter 10, for a detailed treatment of the subject.

11. A fuller treatment of the model with drifting regressors may be found in Mills and Markellos, *Econometric Modelling*, chapter 9.2.
12. The *dw* statistic as a test for cointegration has been analysed by J. Denis Sargan and Alok S. Bhargava, 'Testing residuals from least squares regression for being generated by the Gaussian random walk', *Econometrica* 51 (1983), 153–74, and Bhargava, 'On the theory of testing for unit roots in observed time series', *Review of Economic Studies* 53 (1986), 369–84.
13. This was proposed by Phillips and Bruce E. Hansen, 'Statistical inference in instrumental variables regression with $I(1)$ processes', *Review of Economic Studies* 57 (1990), 99–125: see Mills and Markellos, *Econometric Modelling*, chapter 9.4, for details.
14. See, for example, Phillips and Mica Loretan. 'Estimating long-run economic equilibria', *Review of Economic Studies* 58 (1991), 407–36; Penti Saikkonen, 'Asymptotically efficient estimation of cointegrating regressions', *Econometric Theory* 7 (1991), 1–21; James H. Stock and Mark W. Watson, 'A simple estimator of cointegrating vectors in higher order integrated systems', *Econometrica* 61 (1993), 783–820.

9
Cointegration in Systems of Equations

VARs with integrated variables

9.1 How do non-stationarity and the possible presence of cointegration manifest themselves in systems of equations? Consider again the VAR(p) process of §7.5

$$\mathbf{y}_t = \mathbf{c} + \sum_{i=1}^{p} \mathbf{A}_i \mathbf{y}_{t-i} + \mathbf{u}_t \qquad (9.1)$$

where $\mathbf{y}_t$ and $\mathbf{u}_t$ are both $n \times 1$ vectors,

$$E(\mathbf{u}_t) = \mathbf{0}$$

and

$$E(\mathbf{u}_t \mathbf{u}_s') = \begin{cases} \Omega_p, & t = s \\ \mathbf{0}, & t \neq s \end{cases}$$

From §7.12, (9.1) can be written in lag operator form as

$$\mathbf{A}(B)\mathbf{y}_t = \mathbf{c} + \mathbf{u}_t \qquad (9.2)$$

where

$$\mathbf{A}(B) = \mathbf{I}_n - \sum_{i=1}^{p} \mathbf{A}_i B^i$$

Assuming $p > 1$, the matrix polynomial $A(B)$ can always be written as[1]

$$A(B) = (I_n - AB) - \Phi(B)B(1-B)$$

where

$$A = \sum_{i=1}^{p} A_i$$

and

$$\Phi(B) = \sum_{i=1}^{p-1} \Phi_i B^{i-1}, \quad \Phi_i = -\sum_{j=i+1}^{p} A_j$$

The Φ_i can be obtained recursively from $\Phi_1 = -A + A_1$ as $\Phi_i = \Phi_{i-1} + A_i$, $i = 2,...,p-1$. With this decomposition of $A(B)$, (9.2) can always be written as

$$((I - AB) - \Phi(B)\Delta)y_t = c + u_t$$

or

$$y_t = c + \Phi(B)\Delta y_{t-1} + Ay_{t-1} + u_t$$

An equivalent representation is

$$\Delta y_t = c + \Phi(B)\Delta y_{t-1} + \Pi y_{t-1} + u_t \tag{9.3}$$

where

$$\Pi = A - I_n = -A(1)$$

is known as the **long-run matrix**. The representation (9.3) is the multivariate counterpart of the ADF regression (recall Example 4.1) and we should emphasise that it is a purely *algebraic* transformation of (9.2) as no assumptions about the properties of the y_t vector have been made up to this point.

9.2 Consider now the case where $A = I_n$, so that $\Pi = 0$ in (9.3) and Δy_t follows the VAR($p-1$) process

$$\Delta y_t = c + \Phi(B)\Delta y_{t-1} + u_t \tag{9.4}$$

Setting $A = I_n$ implies that

$$|\Pi| = |A_1 + \ldots + A_p - I_n| = 0 \tag{9.5}$$

Since this is the multivariate extension of the unit root condition of §4.6, the VAR (9.2) is then said to contain *at least one* unit root, y_t is an $I(1)$ process and a VAR in the first differences Δy_t, as in (9.4), is the appropriate specification.

VARs with cointegrated variables

9.3 Note that (9.5) does not necessarily imply that $A = I_n$ and it is this fact that leads to cointegrated VARs. Thus suppose that (9.5) holds, so that the long-run matrix Π is singular and $|\Pi| = 0$, but $\Pi \neq 0$ and $A \neq I_n$. Being singular, Π will thus have reduced rank, say r, where $0 < r < n$. In such circumstances, Π can be expressed as the product of two $n \times r$ matrices β and α, both of full column rank r, so that $\Pi = \beta\alpha'$.

To see why this is the case, note that α' can be defined as the matrix containing the r linearly independent rows of Π, so that Π must be able to be written as a linear combination of α': β is then the matrix of coefficients that are needed to be able to do this. These r linearly independent rows of Π, contained as the rows of $\alpha' = (\alpha_1, \ldots, \alpha_r)'$, are known as the *cointegrating vectors* and Π will contain only $n-r$ unit roots, rather than the n unit roots that it would contain if $\Pi = 0$, which will be the case if $r = 0$.[2]

9.4 Why are the rows of α' known as cointegrating vectors? Substituting $\Pi = \beta\alpha'$ into (9.4) yields

$$\Delta y_t = c + \Phi(B)\Delta y_{t-1} + \beta\alpha' y_{t-1} + u_t$$

The assumption that $\mathbf{y}_t$ is $I(1)$ implies that, since $\Delta\mathbf{y}_t$ must then be $I(0)$, $\boldsymbol{\alpha}'\mathbf{y}_t$ must also be $I(0)$ for both sides of the equation to 'balance'. In other words, $\boldsymbol{\alpha}'$ is a matrix whose rows, when post-multiplied by $\mathbf{y}_t$, produce *stationary* linear combinations of $\mathbf{y}_t$: the r linear combinations $\boldsymbol{\alpha}_1\mathbf{y}_t,\ldots, \boldsymbol{\alpha}_r\mathbf{y}_t$ are all stationary.

9.5 Consequently, if $\mathbf{y}_t$ is cointegrated with cointegrated rank r, then it can be represented as the *vector error correction model* (VECM)

$$\Delta\mathbf{y}_t = \mathbf{c} + \Phi(B)\Delta\mathbf{y}_{t-1} + \boldsymbol{\beta}\mathbf{e}_{t-1} + \mathbf{u}_t \tag{9.6}$$

where $\mathbf{e}_t = \boldsymbol{\alpha}'\mathbf{y}_t$ are the r stationary *error corrections*. This is known as Granger's Representation Theorem and is clearly the multivariate extension and generalisation of (8.8).

9.6 Several additional points are worth mentioning. The parameter matrices $\boldsymbol{\alpha}$ and $\boldsymbol{\beta}$ are not uniquely identified, since for any non-singular $r \times r$ matrix $\boldsymbol{\xi}$, the products $\boldsymbol{\beta\alpha}'$ and $\boldsymbol{\beta\xi}(\boldsymbol{\xi}^{-1}\boldsymbol{\alpha}')$ will both equal Π. If $r = 0$ then we have already seen in §9.2 that the model becomes the VAR(p–1) process (9.4) in the first differences $\Delta\mathbf{y}_t$. If, on the other hand, $r = n$, then Π is of full rank and is non-singular, and $\mathbf{y}_t$ will contain *no* unit roots and will be $I(0)$, so that a VAR(p) in the levels of $\mathbf{y}_t$ is appropriate from the outset.

The error corrections $\mathbf{e}_t$, although stationary, are not restricted to having zero means, so that, as (9.6) stands, growth in $\mathbf{y}_t$ can come about via both the error correction $\mathbf{e}_t$ and the autonomous drift component $\mathbf{c}$. How this constant, and also, perhaps, a trend, are treated is important in determining the appropriate estimation procedure and the set of critical values used for inference.[3]

Estimation of VECMs and tests of cointegrating rank

9.7 Maximum likelihood (ML) estimation of the VECM (9.6) is discussed in many texts and computational routines are available in most econometrics packages. Without going into unnecessary technical details, ML estimates are obtained in the following way. Consider (9.6) again but now written as

$$\Delta\mathbf{y}_t = \mathbf{c} + \sum_{i=1}^{p-1}\Phi_i\Delta\mathbf{y}_{t-i} + \boldsymbol{\beta\alpha}'\mathbf{y}_{t-1} + \mathbf{u}_t \tag{9.7}$$

The first step is to estimate (9.7) under the restriction $\beta\alpha' = 0$. As this is simply a VAR($p-1$) in $\Delta\mathbf{y}_t$, OLS estimation will yield the set of residuals $\hat{\mathbf{u}}_t$, from which is calculated the sample covariance matrix

$$\mathbf{S}_{00} = T^{-1}\sum_{t=1}^{T}\hat{\mathbf{u}}_t\hat{\mathbf{u}}_t'$$

The second step is to estimate the multivariate regression

$$\mathbf{y}_{t-1} = \mathbf{d} + \sum_{i=1}^{p-1}\Xi_i\Delta\mathbf{y}_{t-i} + \mathbf{v}_t$$

and use the OLS residuals $\hat{\mathbf{v}}_t$ to calculate the covariance matrices

$$\mathbf{S}_{11} = T^{-1}\sum_{t=1}^{T}\hat{\mathbf{v}}_t\hat{\mathbf{v}}_t'$$

and

$$\mathbf{S}_{10} = T^{-1}\sum_{t=1}^{T}\hat{\mathbf{u}}_t\hat{\mathbf{v}}_t' = \mathbf{S}_{01}'$$

In effect, these two regressions partial out the effects of $\Delta\mathbf{y}_{t-1},...,\Delta\mathbf{y}_{t-p+1}$ from $\Delta\mathbf{y}_t$ and $\mathbf{y}_{t-1}$, leaving us to concentrate on the relationship between $\Delta\mathbf{y}_t$ and $\mathbf{y}_{t-1}$, which is parameterised by $\beta\alpha'$. α is then estimated by the r linear combinations of $\mathbf{y}_{t-1}$ which have the largest squared partial correlations with $\Delta\mathbf{y}_t$: this is known as a *reduced rank regression*.

9.8 More precisely, this procedure maximises the likelihood of (9.7) by solving a set of equations of the form

$$(\lambda_i\mathbf{S}_{11} - \mathbf{S}_{10}\mathbf{S}_{00}^{-1}\mathbf{S}_{01})v_i = 0 \quad i = 1,...,n \qquad (9.8)$$

where $\hat{\lambda}_1 > \hat{\lambda}_2 > ... > \hat{\lambda}_n$ are the set of eigenvalues and $\mathbf{V} = (v_1, v_2,..., v_n)$ contains the set of associated eigenvectors, subject to the normalisation

$$\mathbf{V}'\mathbf{S}_{11}\mathbf{V} = \mathbf{I}_n$$

Cointegration in Systems of Equations 139

The ML estimate of α is then given by the eigenvectors corresponding to the r largest eigenvalues:

$$\hat{\alpha} = (v_1, v_2, ..., v_r)$$

and the ML estimate of β is consequently calculated as

$$\hat{\beta} = S_{01} \hat{\alpha}$$

which is equivalent to the estimate of β that would be obtained by substituting $\hat{\alpha}$ into (9.7) and estimating by OLS, which also provides ML estimates of the remaining parameters in the model.[4]

9.9 Of course, ML estimation is based upon a known value of the cointegrating rank r and in practice this value will be unknown. Fortunately, the set of equations (9.8) also provides a method of determining the value of r. If $r = n$ and Π is unrestricted, the maximised log likelihood is given by

$$L(n) = K - (T/2) \sum_{i=1}^{n} \log(1 - \lambda_i)$$

where $K = -(T/2)(n(1 + \log 2\pi) + \log|S_{00}|)$. For a given value of $r < n$, only the first r eigenvalues should be positive, and the restricted log likelihood is

$$L(r) = K - (T/2) \sum_{i=1}^{r} \log(1 - \lambda_i)$$

An LR test of the hypothesis that there are r cointegrating vectors against the alternative that there are n is thus given by

$$\eta_r = 2(L(n) - L(r)) = -T \sum_{i=r+1}^{n} \log(1 - \lambda_i)$$

This is known as the **trace statistic** and testing proceeds in the sequence $\eta_0, \eta_1, ..., \eta_{n-1}$. A cointegrating rank of r is selected if the *last* significant statistic is η_{r-1}, which thereby rejects the hypothesis of $n - r + 1$ unit roots in Π. The trace statistic measures the importance of the adjustment coefficients β on the eigenvectors to be potentially omitted. An alternative test of the significance of the largest eigenvalue is

$$\zeta_r = -T\log(1-\lambda_{r+1}) \quad r = 0,1,\ldots,n-1$$

which is known as the **maximal-eigenvalue** or **λ-max statistic**. Both η_r and ζ_r have non-standard limiting distributions that are generalisations of the Dickey–Fuller unit root distributions. The limiting distributions depend on n and on restrictions imposed on the behaviour of the constant and trend appearing in the VECM. For example, if **c** in (9.7) is replaced by $\mathbf{c}_0 + \mathbf{c}_1 t$, then both the ML estimation and testing procedures need to be amended to take into account the presence of a linear trend and the various possible restrictions that could be placed on $\mathbf{c}_0$ and $\mathbf{c}_1 t$.[5]

EXAMPLE 9.1   A simple example of the algebra of VECMs

Let us assume that $p = n = 2$, so that we have a VAR(2) in the variables y_1 and y_2 (intercepts are omitted for simplicity):

$$y_{1,t} = a_{11,1}y_{1,t-1} + a_{12,1}y_{2,t-1} + a_{11,2}y_{1,t-2} + a_{12,2}y_{2,t-2} + u_{1,t}$$

$$y_{2,t} = a_{21,1}y_{1,t-1} + a_{22,1}y_{2,t-1} + a_{21,2}y_{1,t-2} + a_{22,2}y_{2,t-2} + u_{2,t}$$

The various coefficient matrices required for (9.3) are

$$\mathbf{A}_1 = \begin{bmatrix} a_{11,1} & a_{12,1} \\ a_{21,1} & a_{22,1} \end{bmatrix} \quad \mathbf{A}_2 = \begin{bmatrix} a_{11,2} & a_{12,2} \\ a_{21,2} & a_{22,2} \end{bmatrix}$$

$$\mathbf{A} = \mathbf{A}_1 + \mathbf{A}_2 = \begin{bmatrix} a_{11,1} + a_{11,2} & a_{12,1} + a_{12,2} \\ a_{21,1} + a_{21,2} & a_{22,1} + a_{22,2} \end{bmatrix} = \begin{bmatrix} a_{11} & a_{12} \\ a_{21} & a_{22} \end{bmatrix}$$

$$\Pi = \mathbf{A} - \mathbf{I} = \begin{bmatrix} \pi_{11} & \pi_{12} \\ \pi_{21} & \pi_{22} \end{bmatrix} = \begin{bmatrix} a_{11}-1 & a_{12} \\ a_{21} & a_{22}-1 \end{bmatrix}$$

The singularity condition on the long-run matrix Π is

$$|\Pi| = 0 = \pi_{11}\pi_{22} - \pi_{12}\pi_{21}$$

which implies that

$$\Pi = \begin{bmatrix} \pi_{11} & \pi_{12} \\ (\pi_{22}/\pi_{12})\pi_{11} & (\pi_{22}/\pi_{12})\pi_{12} \end{bmatrix}$$

and

$$\beta = \begin{bmatrix} 1 \\ \pi_{22}/\pi_{12} \end{bmatrix} \quad \alpha' = \begin{bmatrix} \pi_{11} & \pi_{12} \end{bmatrix}$$

or, equivalently, using $\xi = \pi_{11}$,

$$\beta = \begin{bmatrix} \pi_{11} \\ (\pi_{22}/\pi_{12})\pi_{11} \end{bmatrix} \quad \alpha' = \begin{bmatrix} 1 & \pi_{12}/\pi_{11} \end{bmatrix}$$

The VECM (9.7) is then

$$\Delta \mathbf{y}_t = -\mathbf{A}_2 \Delta \mathbf{y}_{t-1} + \beta \alpha' \mathbf{y}_{t-1} + \mathbf{u}_t$$

or

$$\begin{bmatrix} \Delta y_{1,t} \\ \Delta y_{2,t} \end{bmatrix} = -\begin{bmatrix} a_{11,2} & a_{12,2} \\ a_{21,2} & a_{22,2} \end{bmatrix} \begin{bmatrix} \Delta y_{1,t-1} \\ \Delta y_{2,t-1} \end{bmatrix} + \begin{bmatrix} \pi_{11} \\ (\pi_{22}/\pi_{12})\pi_{11} \end{bmatrix} \begin{bmatrix} 1 & \pi_{12}/\pi_{11} \end{bmatrix} \begin{bmatrix} y_{1,t-1} \\ y_{2,t-1} \end{bmatrix} + \begin{bmatrix} u_{1,t} \\ u_{2,t} \end{bmatrix}$$

Written equation by equation, this is

$$\Delta y_{1,t} = -a_{11,2}\Delta y_{1,t-1} - a_{12,2}\Delta y_{2,t-1} + \pi_{11}e_{t-1} + u_{1,t}$$

$$\Delta y_{2,t} = -a_{21,2}\Delta y_{1,t-1} - a_{22,2}\Delta y_{2,t-1} + (\pi_{22}/\pi_{12})\pi_{11}e_{t-1} + u_{2,t}$$

$$e_t = y_{1,t} - (\pi_{12}/\pi_{11})y_{2,t}$$

The various π_{rs} coefficients can themselves be expressed in terms of the $a_{rs,i}$ coefficients, $r, s, i = 1,2$, if desired.

EXAMPLE 9.2   A VECM representation of long and short interest rates

We now consider the vector of interest rate *levels*, $\mathbf{y}_t = (RS_t, R20_t)'$. Since lag order criteria statistics are not affected by the presence of possible unit roots in $\mathbf{y}_t$, Table 9.1 shows that $p = 3$ is a suitable choice for the order of the levels VAR. Since $n = 2$, setting $r = 0$ would imply that there was no cointegration and the representation would be that found in Example 7.1, a VAR(2) in the differences ΔRS and $\Delta R20$. If $r = 1$ there will be a single cointegrating vector with the error correction $e_t = \alpha_1 RS_t + \alpha_2 R20_t + \alpha_0$, where a constant is allowed. If $r = 2$ then there are no unit roots and the levels VAR(3) is appropriate.

Including c in (9.7), allowing a constant in the cointegrating vector and estimating by ML obtains the eigenvalues $\lambda_1 = 0.0201$ and $\lambda_2 = 0.0018$, using which the trace statistics $\eta_0 = 16.67$ and $\eta_1 = 1.39$ and maximum eigenvalue statistics $\xi_0 = 15.28$ and $\xi_1 = 1.39$ are calculated. The η_0 and ξ_0 statistics reject the null hypothesis of $r = 0$ in favour of $r > 0$ at marginal significance levels of 0.033 and 0.035 respectively, but the η_1 and ξ_1 statistics, which by definition are equal in this example, cannot reject the null of $r = 1$ in favour of $r = 2$. We are thus led to the conclusion that RS and $R20$ are indeed cointegrated.

Table 9.1   Order determination statistics for $\mathbf{y}_t = (RS_t, R20_t)'$

p	logL	LR(p,p–1)	MAIC	MBIC
0	–3508.4	–	9.336	9.348
1	–533.9	5925.20	1.436	1.473
2	–468.8	129.27	1.273	1.335*
3	–462.7	12.09*	1.268*	1.354
4	–462.2	0.94	1.277	1.388

$LR(p,p-1) \sim \chi_4^2$ $\chi_4^2(0.05) = 9.49$

ML estimation of the implied VECM obtained the following estimates, written in individual equation form for convenience

$$\Delta RS_t = 0.226\ \Delta RS_{t-1} + 0.039\ \Delta RS_{t-2} + 0.272\ \Delta R20_{t-1}$$
$$(0.041)\qquad\quad (0.040)\qquad\quad (0.062)$$
$$-\ 0.079\ \Delta R20_{t-2} + 0.026\ e_{t-1} + \hat{u}_{1,t}$$
$$(0.063)\qquad\quad\ \ (0.008)$$

$$\Delta R20_t = -\ 0.012\ \Delta RS_{t-1} + 0.020\ \Delta RS_{t-2} + 0.311\ \Delta R20_{t-1}$$
$$(0.027)\qquad\quad (0.026)\qquad\quad (0.041)$$
$$-\ 0.137\ \Delta R20_{t-2} - 0.003\ e_{t-1} + \hat{u}_{2,t}$$
$$(0.041)\qquad\quad\ \ (0.005)$$

$$e_t = R20_t - 1.053\ RS_t - 0.827$$
$$\qquad\ \ (0.132)\qquad\ \ (0.997)$$

The error correction has been normalised by setting $\alpha_2 = 1$; this helps to identify the cointegrating vector. Examination of the estimated coefficients suggests two important potential restrictions: the error correction in the $\Delta R20$ equation is insignificantly different from 0 and the coefficient on RS in the cointegrating vector is insignificantly different from unity. The first would imply that $R20$ is exogenous in the long-run, while the second implies that the error correction can be interpreted as being the deviation from the spread. Imposing $\beta_1 = 0$ and $\alpha_1 = 1$ produces

$$\Delta RS_t = 0.226\ \Delta RS_{t-1} + 0.038\ \Delta RS_{t-2} + 0.272\ \Delta R20_{t-1}$$
$$(0.041)\qquad\quad (0.040)\qquad\quad (0.062)$$
$$-\ 0.079\ \Delta R20_{t-2} + 0.030\ e_{t-1} + \hat{u}_{1,t}$$
$$(0.063)\qquad\quad\ \ (0.008)$$

$$\Delta R20_t = -\ 0.013\ \Delta RS_{t-1} + 0.019\ \Delta RS_{t-2} + 0.312\ \Delta R20_{t-1}$$
$$(0.027)\qquad\quad (0.026)\qquad\quad (0.041)$$
$$-\ 0.137\ \Delta R20_{t-2} + \hat{u}_{2,t}$$
$$(0.041)$$

$$e_t = R20_t - RS_t - 1.184$$
$$(0.473)$$

The two restrictions are not rejected by an LR test and thus confirm that $R20$ is exogenous; in fact, since the coefficients on the lagged ΔRS terms in the $\Delta R20$ equation are both insignificant, $R20$ is therefore completely exogenous and could be represented as a univariate AR(2) process in the differences. The spread is found to be the cointegrating vector, so that the error correction is the deviation of the spread from its equilibrium value of 1.184. Figure 9.1 shows the error correction and it is clearly seen to be stationary.

The VECM may thus be interpreted as saying that a shock to the gilt market, leading to a change in the long interest rate, induces a change in short interest rates in the bond market, but a shock to the bond market does not produce any change in the long rate.

Nevertheless, there is a long-run equilibrium in which the spread between long and short rates is around 1.2%. When interest rates move so much that the spread deviates substantially from this equilibrium, factors come into play to ensure that this deviation is eradicated as the equilibrium error corrects itself. However, the adjustment coefficient of 0.030 ensures that this correction is slow, so that equilibrium errors are very persistent, and this is consistent with the findings from the univariate analysis of the spread given in Examples 2.2 and 4.1.

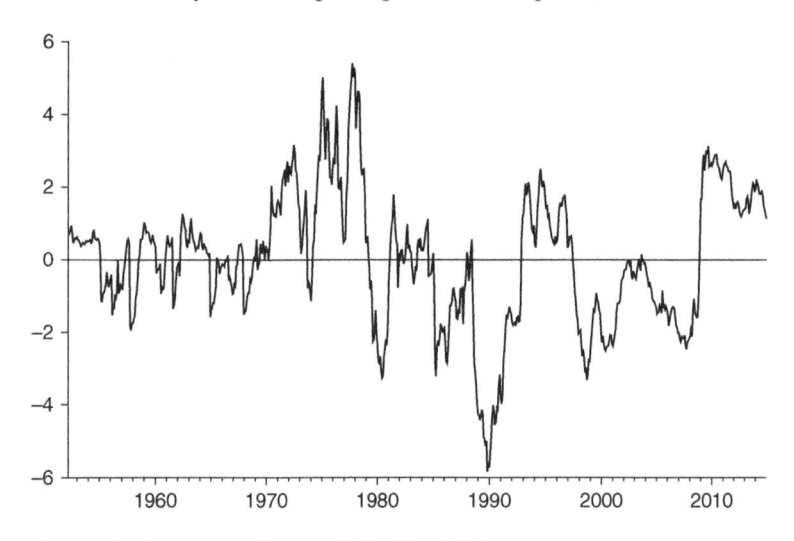

Figure 9.1 Error correction $e_t = R20_t - RS_t - 1.184$

EViews Exercises

9.10 To produce the results contained in Example 9.2 open r20 and rs in page Ex_2_2 as a group. Table 9.1 can be constructed by following the procedure of §7.16. The trace and max-eigenvalue statistics for cointegration are obtained by clicking *View/Cointegration Test/Johansen System Cointegration Test* and changing 'Lag intervals' to 1 2.

To produce the VECM estimates, click *Proc/Make Vector Autoregression...*, select 'Vector Error Correction' as the 'VAR Type', click *Cointegration* and choose option 2 as the 'Deterministic Trend Specification'. To estimate the restricted VECM click *Estimate/VEC Restrictions* and in the 'VEC Coefficient Restrictions' box check 'Impose Restrictions' and insert

```
b(1,1)=1, b(1,2)=-1, a(1,1)=0
```

The test for the validity of these restrictions is provided and is clearly insignificant. To obtain the error correction of Figure 9.1 click *View/Cointegration Graph*.

Impulse responses and variance decompositions for the VECM may be obtained in exactly the same way as for the VAR in §7.17.

Notes

1. This representation is the matrix equivalent of the univariate AR representation of §4.6. Again it is most clearly seen for $p = 2$:

    $A(B) = I - A_1B - A_2B^2$
    $= I - A_1B - A_2B + A_2B - A_2B^2$
    $= I - (A_1 + A_2)B + A_2B(1 - B)$
    $= I - AB - \Phi_1B(1 - B)$

    since here $A = A_1 + A_2$ and $\Phi_1 = -A_2$.

2. For detailed text book treatments of this argument, see Anindya Banerjee, Juan Dolado, John W. Galbraith and Hendry, *Co-integration, Error-Correction, and the Econometric Analysis of Non-stationary Data* (Oxford University Press, 1993) and Søren Johansen, *Likelihood-Based Inference in Cointegrated Vector Autoregressive Models* (Oxford University Press, 1995).

3. Mills and Markellos, *Econometric Modelling*, chapter 9.5, discuss how to deal with the inclusion of both constants and trends into VECMs.

4. This procedure can be straightforwardly adapted when a linear trend is included in (9.7) and when various restrictions are placed upon the intercept and trend coefficients. This involves adjusting the first and second step regressions to accommodate these alterations.
5. See, for example, Johansen, *Likelihood-Based Inference*, chapters 6 and 15, for extended discussion.

10
Extensions and Developments

Seasonality, Non-linearities and Breaks

10.1 As we emphasised in Chapter 1, this text book is very much an introduction to time series econometrics: consequently, some topics have not been covered either because they are too peripheral to the main themes or are too advanced. Three areas not covered, but which nevertheless can be important when analysing time series data, are seasonality, non-linearities and breaks.

10.2 Many economic time series have seasonal patterns and, although seasonal adjustment procedures have been developed over many years and form the basis for 'agency' approaches to adjusting seasonal data prior to their analysis or, indeed, publication, there are many circumstances where it is preferable to incorporate seasonal patterns directly into the modelling framework.

The most notable univariate approach is Box and Jenkins' extension of ARIMA models to incorporate seasonal patterns using a multiplicative modelling framework. This is best seen in terms of an ARIMA(0,1,1) model for monthly data. Here the observed series is characterised by the seasonal process

$$(1 - B^{12})x_t = (1 - \Theta B^{12})\alpha_t$$

whose error α_t follows the non-seasonal process

$$(1 - B)\alpha_t = (1 - \theta B)a_t$$

The two processes may then be combined to produce the multiplicative ARIMA(0,1,1)×(0,1,1)$_{12}$ model

$$(1-B)(1-B^{12})x_t = (1-\theta B)(1-\Theta B^{12})a_t \tag{10.1}$$

Box and Jenkins found that this model fitted a time series of international airline passenger numbers extremely well and it has since become known as the 'airline model'. An extension of the identification procedures outlined in Chapters 2 and 3 to non-stationary seasonal time series is straightforward to apply.[1]

The presence of the multiplicative difference filter

$$(1-B)(1-B^{12}) = (1-B)^2(1+B+B^2+\ldots+B^{11})$$

in (10.1) shows that the airline model assumes that x_t contains a set of unit roots, so that the seasonal pattern is both stochastic and non-stationary. There are seasonal unit root tests available with which to test this assumption and, indeed, there are further modelling techniques that will test whether seasonality is stochastic, and so needs to be modelled by some form of ARMA process, or whether it is deterministic, in which case it may be represented as a set of seasonal dummies.[2]

10.3   Recent years have seen great advances in non-linear time series modelling, most notably models that attempt to capture *regime switching*, such as the SETAR (self-exciting threshold autoregressive), STAR (smooth transition autoregressive) and Markov switching processes. Neural networks and chaotic processes have also proved popular for modelling financial time series.[3]

Although not strictly non-linear, long memory processes have become a feature of modelling financial time series. Long memory is associated with an autocorrelation function that declines hyperbolically, so that the decline is slower than the exponential decline of a stationary process but faster than the linear decline associated with an $I(1)$ process. Long memory can be characterised by the use of *fractional differencing* through the operator

$$\Delta^d = (1-B)^d = 1 - dB + \frac{d(d-1)}{2!}B^2 - \frac{d(d-1)(d-2)}{3!}B^3 + \ldots$$

Extensions and Developments 149

where d is now allowed to take *any* value greater than -1, not just integers: $\Delta^d x_t = a_t$ then defines ***fractional white noise*** and, if $|d| < 0.5$, x_t is stationary and invertible but will exhibit long memory. If a_t is autocorrelated it may be modelled as an ARMA process, thus leading to the AR-***fractionally integrated***-MA, or ARFIMA, process.[4]

10.4 Although in §4.8 we contrasted DS and TS processes, a great deal of interest has recently focused on a third type of process, the ***segmented*** (or ***breaking***) ***trend model***, in which the series evolves as stationary deviations around a linear trend that 'breaks' in one or more places. Such models allow shocks to the series to be typically transitory but occasionally permanent when a break occurs (on this perspective a random walk is a breaking trend model with the breaks occurring *every* period).

Under a breaking trend model, unit root testing procedures need to be amended and whether the timings of the breaks are known (exogenously determined breaks) or whether they are unknown and must themselves be estimated (endogenous breaks) becomes an extremely important question since, for example, critical values of test statistics are affected. Breaks may also be introduced into cointegration relationships, thus leading to the concept of ***temporary cointegration***, in which the cointegrating relationship can be switched on or off depending, for example, on the nature of the policy regime in place.[5]

Unobserved Component Models and Trend and Cycle Extraction

10.5 Seasonal adjustment procedures typically have underlying them an unobserved components (UC) decomposition in which the observed series x_t is represented as either the sum or product of the unobserved non-seasonal, n_t, and seasonal, s_t, components: $x_t = n_t + s_t$ is an additive decomposition while $x_t = n_t \times s_t$ is a multiplicative decomposition, the latter implying that there is an additive decomposition of the logarithms. ***Signal extraction*** procedures thus attempt to estimate the unobserved components from a realisation of x_t. When there are assumed, typically ARIMA, models for n_t and s_t then we have model-based seasonal adjustment, for example, but general filters can be used to extract trend and cyclical components from extended decompositions. The Hodrick–Prescott trend is a popular example of a filter of this type, but several others have also been proposed.[6]

UC models can often be given a *state space representation* within which estimation of their parameters, forecasting of future values, and estimation of the components (known as smoothing) can all be performed using the *Kalman filter*.[7]

Common Trends and Features and Co-breaking

10.6 The presence of cointegration is often thought of in terms of there being *common stochastic trends*: if a set of n time series have r cointegrating vectors then they equivalently have $n-r$ common stochastic trends. The concept of common trends may be extended to that of *common features*: for example, if two series individually contain seasonal or cyclical patterns but a linear combination of them does not, then these series contain common seasonal or cyclical features.[8]

One important manifestation of a common feature is when there are common structural breaks across a set of series, a phenomenon known as *co-breaking* and which has clear links with models of cointegration in the presence of structural breaks or regime shifts.[9]

Generalisations of Cointegration and VECMs

10.7 Several generalisations of VECM modelling are now available (it is now often termed *cointegrated VAR* (CVAR) modelling). The possibility of cointegrated $I(2)$ series may be entertained, although the analysis is considerably more complicated than the standard approach when the data are just $I(1)$.[10]

10.8 A variety of non-linear extensions to cointegration have been proposed. These typically take one of two forms; either a linear cointegration vector is allowed to enter as a non-linear error correction, or the cointegrating relationship itself is specified to be non-linear. A popular approach is to use threshold-type processes for modelling non-linear error corrections, so that only large errors from equilibrium, those above some threshold, are corrected.[11]

Extensions and Developments 151

Notes

1. Mills, *Foundations*, chapter 14, provides discussion of the historical evolution of seasonal adjustment procedures and also sets out the Box–Jenkins approach to seasonal ARIMA modelling, the original development being contained in chapter 9 of their *Time Series Analysis*.
2. The initial paper on testing for seasonal unit roots is Sven Hylleberg, Engle, Granger and Byung Sam Yoo, 'Seasonal integration and cointegration', *Journal of Econometrics* 44 (1990), 215–38. How to distinguish between different forms of seasonal patterns is considered in Mills and Alessandra G. Mills, 'Modelling the seasonal patterns in UK macroeconomic time series', *Journal of the Royal Statistical Society, Series A* 155 (1992), 61–75.
3. A useful text book on non-linear models is Philip Hans Franses and Dick van Dijk, *Non-linear Time Series Models in Empirical Finance* (Cambridge University Press, 2000). Timo Teräsvirta, 'Univariate nonlinear time series models', chapter 10 of Mills and Patterson, *Palgrave Handbook, Volume 1*, 396–424, is a recent survey of the area.
4. Long memory is also referred to as **persistency** and is sometimes known as the **Hurst effect**, after the hydrologist Harold E. Hurst, who encountered this phenomenon when analysing records of river flow for the Nile.
   The notion of fractional differencing seems to have been proposed contemporaneously and independently by J.R.M. Hosking, 'Fractional differencing', *Biometrika* 68 (1981), 165–76, and Granger and Roselyn Joyeux, 'An introduction to long memory time series models and fractional differencing', *Journal of Time Series Analysis* 1 (1981), 15–29. A recent survey is Luis A. Gil-Alana and Javier Hualde, 'Fractional integration and cointegration: an overview and an empirical application', chapter 10 of Mills and Patterson, *Palgrave Handbook of Econometrics, Volume II: Applied Econometrics*, 434–69 (Cambridge University Press, 2009).
5. Testing for unit roots in breaking trend models was first considered in Perron, 'The Great Crash, the oil price shock, and the unit root hypothesis', *Econometrica* 57 (1989), 1361–401. The literature on modelling breaks has since grown enormously: see, for example, Perron, 'Dealing with structural breaks', chapter 8 of Mills and Patterson, *Palgrave Handbook, Volume 1*, 278–352. Temporary cointegration was introduced by Pierre Siklos and Granger, 'Temporary cointegration with an application to interest rate parity', *Macroeconomic Dynamics* 1 (1997), 640–57, and various tests of breaks in cointegrating relationships have been proposed.
6. Robert J. Hodrick and Edward C. Prescott, 'Postwar U.S. business cycles: an empirical investigation', *Journal of Money, Credit and Banking* 29 (1997), 1–16. An introductory text book on the modelling of trends and cycles is Mills, *Modelling Trends and Cycles in Economic Time Series* (Palgrave Macmillan, 2003), while a more detailed treatment is provided by D. Stephen G. Pollock, 'Investigating economic trends and cycles', chapter 6 of Mills and Patterson, *Palgrave Handbook of Econometrics, Volume II*, 243–307.

7. Textbook treatments are provided by Andrew C. Harvey, *Forecasting, Structural Time Series Models and the Kalman Filter* (Cambridge University Press, 1989) and Durbin and Siem Jan Koopman, *Time Series Analysis by State Space Methods*, 2nd edition (Oxford University Press, 2012).
8. This common trends interpretation is emphasised by Stock and Watson, 'Testing for common trends', *Journal of the American Statistical Association* 83 (1988), 1097–107. Testing for and modelling common features were first introduced in Engle and Sharon Kozicki, 'Testing for common features', *Journal of Business Economics and Statistics* 11 (1993), 369–80, and is generalised and surveyed in Farshid Vahid, 'Common cycles', chapter 16 of Mills and Patterson, *Palgrave Handbook of Econometrics, Volume I*, 610–30.
9. See Hendry and Michael Massmann, 'Co-breaking: recent advances and a synopsis of the literature', *Journal of Business Economics and Statistics* 25 (2007), 33–51.
10. See Johansen, 'A statistical analysis of cointegration for $I(2)$ variables', *Econometric Theory* 11 (1995), 25–59 and, for a detailed empirical application, Katerina Juselius, 'The long swings puzzle: what the data tell when allowed to speak freely', chapter 8 of Mills and Patterson, *Palgrave Handbook of Econometrics, Volume II*, 349–84.
11. A convenient survey of non-linear cointegration models is provided by Mills and Markellos, *Econometric Modelling*, Chapter 10.2.

Index

adequacy, of models 32–3
Akaike's information criteria
  (AIC) 24–5, 102
augmented Dickey–Fuller tests 63–4
autocorrelation function (ACF) 7,
  11, 16
  partial autocorrelation (PACF) 21
  sample autocorrelation function
    (SACF) 28
  sample partial autocorrelation
    function (SPACF) 28
autocovariance 7
autoregressive fractionally
  integrated moving average
  (ARFIMA) model 149
autoregressive heteroskedastic
  (ARCH) models 74–83
  development of generalised ARCH
    processes 76–7
  testing for presence of ARCH
    errors 77–8
autoregressive integrated
  moving average (ARIMA)
  model 45–55
  forecasting using 85–9
  modelling dollar/sterling
    exchange rate 53–4
  modelling UK spread as integrated
    process 52–3
  non-stationary processes
    and 45–55
  seasonal modelling 147–8
autoregressive moving average
  (ARMA) model 25–35
  model building procedures
    28–35
  returns on *FTA All Share*
    index 33–5
  returns on *S&P 500* 29–30
  UK interest rate spread 31–3

sample autocorrelation and
  partial autocorrelation
  functions 28–9
autoregressive processes 10–13,
  16–22

backshift operator 10

causality 100–4
  tests of 103–4
co-breaking 150
cointegrated processes 122–31
  estimating cointegrating
    regressions 129–31
  testing for cointegration
    127–9
common trends and features 150
component GARCH model 80
conditional expectation 44
Cramer's Rule 22
conditional standard deviation 73

deterministic trends 52
Dickey–Fuller regression 63
Dickey–Fuller tests 61–5
  extensions 63–4
difference stationary processes 65
dynamic regression model 98–9

equilibrium error 122
equilibrium, long-run 122
ergodicity 7
error correction model 122
  estimation of VECMs and tests of
    cointegrating rank 137–40
exchange rates
  GARCH models of 81–3
  modelling dollar/sterling
    exchange rate 53–5
exogenous variables 98–100

expectations 5–6
    conditional 44
explosive processes 44
exponential GARCH model 79
feedback 101
first-difference operator 42
forecast errors 89–90
forecasting
    using autoregressive integrated
        moving average (ARIMA)
        model 85–9
    using autoregressive conditional
        heteroskedastic (ARCH)
        model 94–5
    using trend stationary (TS)
        model 92–3
fractional differencing 149
fractional white noise 149

generalised ARCH (GARCH)
    model 76–7
    exchange rates 81–3
    modification of generalised ARCH
        processes 78–81
    non–linear GARCH 80
generalised impulse responses 106
GJR model 80
Granger–causality 100–4
    instantaneous 100
    tests of 103–4

Hodrick–Prescott (H–P) trend 150
homogenous
    non–stationarity 49–50

impulse response function
    105–6
innovation accounting 105–6
innovations 8
integrated process 46
    determining order of integration
        of a time series 58–60
    modelling UK interest rate spread
        as integrated process 52–3
    interest rates 31
    spread 31–3

autoregressive moving average
    (ARMA) model 31–3
    modelling UK spread as
        integrated process 52–3
    testing for more than one unit
        root 68

kurtosis 74

lag operator 10
Lagrange multiplier (LM) test of
    ARCH 78
least squares 31
likelihood ratio (LR) test
    for VAR lag order 102
    for cointegration 139–40
linear stochastic process 6
long memory 148–9
long-run equilibrium 122

maximum likelihood (ML)
    estimation 77, 137–9
memory 8
    long memory 148–9
minimum mean square error
    (MMSE) forecast 85
moving average processes 13–25
    autoregressive moving average
        (ARMA) model 25–7

non-linear GARCH processes 80
non-linear generalisations of
    cointegration and error
    correction models 150
non-stationary processes,
    autoregressive integrated
    moving average (ARIMA)
    model 45–55
non-stationarity in mean 41
non-stationarity in variance 73–4
normality 5–7

over-differencing 59–60
overfitting 33

partial autocorrelation function
    (PACF) 21

Index 155

portmanteau (Q) statistic 29, 32
power ARCH (PARCH) model 79
probability, stochastic processes
    and 5
product process 73

quasi-maximum likelihood
    (QML) 81

random walk 45–6
    realisation 5–6
regression models
    cointegrated processes 122–31
    estimating cointegrating
        regressions 129–31
    testing for cointegration 127–9
    dynamic regression 98–9
    integrated time series 114–22
    multivariate dynamic linear
        regression model 99
    spurious regression 114–22
    variance decompositions
        and innovation
        accounting 104–110
    vector autoregressions
    determining the lag order of a
        VAR 101–3
    with integrated variables 134–6
    with cointegrated
        variables 136–7
    test of Granger-causality 103–4

sample autocorrelation function
    (SACF) 28
sample partial autocorrelation
    function (SPACF) 28
Schwarz's information criterion
    (BIC) 34–5, 102
seasonality 147–8
signal extraction 149
spurious regression 114–22
Standard & Poor 500 stock return,
    autoregressive moving
    average (ARMA) model 29–30
standardised process 73
state space representation 150
stationarity

definition 6–7
difference stationary
    processes 65–8
strict 7
trend stationary processes
    65–8
weak 7
stochastic process 5
    linear 6
    realisation 5–6
    stationarity 6–7
stochastic trends 52

threshold ARCH model 80
time series
    determining the order of
        integration of a time
        series 58–60
    forecasting using autoregressive
        integrated moving average
        (ARIMA) model 85–9
    regression models
        with integrated time
            series 114–22
    unit root tests 61–5
unobserved component (UC)
    models 149–50
trends
    deterministic trends 52
    Hodrick–Prescott (H–P) trend 150
    trend stationary processes 65–8
    segmented (or breaking)
        trend 149

unit roots 61–5
    tests of 61–5
    augmented Dickey–Fuller
        test 63–4
    more than one unit root
        68–9
univariate linear stochastic model
    autoregressive moving average
        (ARMA) model 25–35
    model building
        procedures 28–35
    non-stationary processes and
        ARIMA model 45–55

univariate linear stochastic model
  – *continued*
  sample autocorrelation and partial autocorrelation functions 28
  autoregressive processes 10–13, 16–22
  decomposing time series 149–50
  unobserved component (UC) models 149–50
  determining the order of integration of a time series 58–60
  ergodicity 6
  fractional integration and long memory 149
  general AR and MA processes 16–25
  linear stochastic processes 6
  moving average processes 13–25
  realisations 5–6
  stationarity 6–7
  stochastic process 5
  testing for unit root 61–5
    extensions to Dickey–Fuller test 63–4
    more than one unit root 68–9
    trend stationarity (TS) versus difference stationarity (DS) 65–8

vector autoregressions (VARs) 99–100
  causality 100–4
  common trends and features 150
  determining the lag order of a VAR 101–3
  with integrated variables 134–6
    cointegrated variables 136–7
    estimation of VECMs and tests of cointegrating rank 137–40
  tests of Granger causality 103–4

white noise process 8
  fractional white noise 149
Wold's decomposition 8, 42

Yule–Walker equations 22

Lightning Source UK Ltd.
Milton Keynes UK
UKOW06n1824251015
261376UK00008B/47/P